Experiencing Mathematics Through Investigations

Ideas from Fifty Years of *Mathematics in School*

Edited by Chris Pritchard & John Berry

MATHEMATICAL ASSOCIATION

Mα

Supporting mathematics in education

THE MATHEMATICAL ASSOCIATION

Experiencing Mathematics Through Investigations

ISBN: 978-1-911616-10-8

FOREWORD

Exploration is what doing mathematics is all about. When we take an idea for a walk, as the articles gathered together in this book do in so many delightful ways, it empowers us, it helps us see ourselves as mathematicians, whatever stage of education we may be at. The work of educators in developing investigations that welcome learners of all ages, and all levels of ability and mathematical maturity, is therefore vitally important.

In this book, Chris Pritchard and John Berry present a variety of investigations from the first fifty years of that most excellent Mathematical Association (MA) publication, *Mathematics in School*. I was honoured to be asked to write the foreword, because I come from an MA family. My parents, Pat and Martin Perkins, were mathematics schoolteachers and textbook authors who were actively involved in the MA. As children, my sister and I knew that every Easter holiday would involve a stay with grandma while Mum and Dad disappeared to whatever an "MA conference" was. Before I knew I was a mathematician, I knew I liked pattern-spotting and playing with shapes and numbers: that age-old child's complaint "I'm bored", in the Perkins household, would regularly be addressed by giving me a mathematical idea to mess around with. One holiday, aged about eight or nine, I spent a couple of days, at the instigation of a passing parent, happily exploring which throws were the most likely with two, three, or more, dice. Another time "give me a puzzle" led to me making tables of symmetries, resulting in investigations about what these tables could look like – group theory by stealth. At about twelve, having learnt Pythagoras' Theorem in school, Mum told me about Pythagorean triples and sent me off to play with them. I still remember the feeling of excitement when I realised how to make a triple starting with any odd number from 3 up, and the next day how to tweak my algorithm to work for even numbers. That excitement of discovery, allied with the pleasure of coming up with a neat explanation of why the thing works, is why I became a mathematician.

I went on to study mathematics at university, and (following a PhD in that same group theory I'd encountered years before disguised as a fun investigation about symmetry) became first a lecturer and ultimately a Professor of Mathematics at Birkbeck as well as being the current Gresham Professor of Geometry. I am one of the incredibly fortunate people whose job is also their hobby. I love playing with mathematical ideas, and I love talking to people about how wonderful mathematics is. I've had lots of articles published in mathematics research journals, but I'm not sure I've ever been quite as proud of any of them as my teenage self was to see a short piece I'd written appear in *Mathematics in School*.

All this is to say that my own experience of investigative play in mathematics was crucial to my development as a mathematician. Looking at the ideas for investigations in this book, I was struck by the variety of problems and puzzles that are covered. From happy numbers to fraudulent dissection puzzles, from counting bubbles to paper weaving, there's something for every taste. I have certainly spent several happy hours exploring some of these ideas. Even where the articles are about familiar results, like Pick's Theorem, there are new approaches and different avenues to follow. I only wish this book had been around in 2020 when, after nearly 20 years teaching in higher education, I, like many other parents, suddenly became a primary school teacher by default, during the COVID-19 lockdown. Some of my happiest times during those difficult months were in "Mummy's maths hour", where my then nine-year old and I explored everything from lattices to cryptography, and we got to experience the joy of mathematical discovery together. I only wish that her idea, to give symmetries names based on the letters of the alphabet they leave invariant, would catch on. (An "Emma" symmetry is reflection in the horizontal, for example.) I emerged from it all with my already considerable respect for schoolteachers increased to something like open hero-worship.

Mathematics in School is a wonderful publication. This book is the first time that so many articles from across its fifty volumes have been brought together in one place. The aim of the editors is to show the incredibly rich variety of contributions, and they succeed magnificently. I hope we shall not have to wait fifty more years for another volume.

Sarah Hart

Gresham Professor of Geometry
Professor of Mathematics, Birkbeck (University of London)
Proud author of 'Investigating four-dimensional figures', *Mathematics in School* (1993)

CONTENTS

Chapter 3: Number Chains

Chapter 4: Geometric Counting

Chapter 5: Number Bases and the Fibonacci Series

Chapter 6: Miscellaneous Investigations

1   *The Investigative Experience*

1.0 Introduction

The 1980s was an exciting decade to be involved in the teaching and learning of Mathematics. Previously, school mathematics was seen as a mixture of 'board taught' methods and 'word problems'. Then phrases such as 'solving real problems with mathematics' began to appear regularly in articles. This 'mathematical modelling' process for solving problems was exemplified by the Open University's 'seven box' modelling diagram.

And, of course in 1982, we saw the publication of *The Cockcroft Report: Mathematics Counts*. There were 810 paragraphs in this comprehensive and influential report of which the most famous and probably most quoted was paragraph 243 which included

> **'Mathematics teaching at all levels should include**
> **opportunities for ... investigational work.'**

Investigations had arrived on the stage of mathematics education, together with the associated discussion and controversy.

We begin this first chapter with an excellent article by David Wells which appeared in January 1985. David has published regularly in *Mathematics in School* and his articles often tackle controversial issues in the teaching and learning of mathematics, so it is appropriate that this is where we begin!

What the Investigation initiative did was to move learning mathematics from a mostly 'learning by watching' to an opportunity to 'learn by exploring'. We see that this was the intention of 'opportunities for investigational work.'

We guess that many teachers had their favourite investigation or collection of investigations. 'Frogs' was certainly well used by both editors. It provided opportunities for the important mathematical processes of 'simplify, generalise, verify'. One of us can still picture a playground of 50 primary children working their way through the 'hops and slides' as the children gradually changed sides. As an observing teacher commented: "they are such a mixed ability group but look at them working together and having fun". This is the essence of many investigational tasks – they are fun and most learners can have a go.

But is 'Frogs' a real problem or an investigation? We leave it to your judgement!

We follow David's article with five quite contrasting investigations. They make interesting reading after the ideas expressed in the first article, 'Problems, investigations and confusion'.

1.1 David Wells on 'Problems, investigations and confusion'
Vol. 14, no. 1 (January 1985), pp 6-9; extract

[Are] problems ... especially linked to applications? Do not questions in class raise problems? What is supposed to be the difference between problems and investigations? Let's look at some opinions.

Being a mathematician

> In the set of real integers, 2 and 13 are primes, having no divisors except themselves and unity. However, in the set of complex numbers they are not prime; $2 = (1 - i)(1 + i)$ and $13 = (3 - 2i)(3 + 2i)$. Are there any prime complex integers?

This is an early reference from 1969: "It is the aim of this project to develop sixth-form work in mathematics in which a major part of the activity is the investigation by individual pupils of substantial open problems ... formulating problems, solving them, extending and generalising them ..." It concludes, "The kind of work exemplified above might be described as research-type activity at the pupils' level" (1). This idea is emphasised elsewhere in the same *Bulletin*: one aim of investigations is "to give students experience of doing mathematics ... The free investigations (in which there is no constraint towards finding a standard result, and the investigation may be taken in any desired direction) seem to me to develop insights into the nature of mathematical work which are not developed by other kinds of activity."

The excellent idea that children should be doing mathematics, being young mathematicians, lies behind all efforts to promote investigations. But is this identification of a special kind of activity, to be called a (free) investigation justified? The next quote, from the same project one year later, suggests that that claim is based on as idiosyncratic an interpretation of 'problem' as Cockcroft displays. After describing exercises as short questions designed to reinforce learning, it continues: "Problems: These are more involved mathematical situations, where several steps, and often the utilisation of a variety of techniques and methods are necessary to their solution ... Investigations: These are more open situations than those described under the previous headings, and may require considerable exploratory work before any conclusions become apparent. It may well not be clear for some time which of several directions to take, and powers of generalisation are called for in a way that is not required in most closed exercises. The student will often have to decide, on the basis of his own developing mathematical judgement, when a suitable conclusion – or dead end – has been reached" (2).

This account of investigations is a fair description of professionals doing mathematics. However, the activity described under *Investigations* is what research mathematicians call *problem solving*, while the description of problem solving describes only the routine solution of a problem using established techniques, that is, it also associates *problem solving* with the *application of mathematics*. It does not describe what research mathematicians mean when they talk about problems. The fact is that professionals do not use the word 'investigation' as it is used by mathematics teachers; they do not separate problems in general from some special 'open' kind of problem called something else, and they approach problems as children are supposed to approach investigations. Thus Halmos writes, "I do believe that problems are the heart of mathematics", Dieudonné that "the history of mathematics shows that a theory almost always originates in the efforts to solve a specific problem" (3). No mention of investigations.

When professionals publish books of problems, they call them by titles such as *Problems in Intuitive Mathematics* (a collection of unsolved problems compiled by Richard Guy). Hilbert in 1900 presented his famous 23 problems to the mathematical world. While the verb 'investigate' might well be used (though it seldom is, in fact) to describe what mathematicians have done to them, they are still called problems, not investigations.

Surely this is extraordinarily curious! That an effort to get pupils to behave like mathematicians should be combined with an interpretation of the word 'problem' which they would not recognise, the introduction of the noun 'investigation' which they do not use, and hence the introduction of a distinction which they never make. Surely it would be more logical to decide that if pupils are going to *do* mathematics, they should start to use the language of the professional, and in particular stop thinking of a problem as merely a difficult exercise, and start to recognise the subtlety and possibilities of real problems, and to learn through their own experience, with their teacher's help, to appreciate those features of professional problem solving, such as generalisation, which my first quote mentioned.

> The government intends to issue a new book of postage stamps to the value of £10. What stamps should the book contain?

Open is in the mind

Professionals also do not describe problems as 'open' in the sense of open-ended, or open to exploration, or with open beginnings. They use the word to describe unsolved problems but that is all. Why this difference? I suggest it is because professionals take for granted that all problems are open to exploration, and quite probably in other senses too, but this openness, except in a very superficial sense, is not a feature of the problem as presented at all, but of the solver's developing relationship with the problem as he tackles it. The superficial sense in which the presentation of a problem may be open is easy to illustrate: in this case the student is simply instructed to investigate: "If you had a 3-litre jug and a 5-litre jug, how could you use them to measure 4 litres? Investigate other problems like this." (4)

The first sentence presents a well-known problem. The second tells the students to consider other problems of the same kind, in effect to take an attitude, to be curious, to want to know what happens if the initial numbers were varied, to wonder what would happen if … But professionals, of course, take just these possibilities, just these attitudes for granted. They apply them to all their problems, because that is how you make progress with one problem, how you progress from one problem to another, how you do research. At the same time, they are aware that their own interests are idiosyncratic, and that they do not find all problems equally fascinating or equally easy to make progress on, (a point I shall return to in relation to children). Their background knowledge may not fit them to easily tackle a particular problem. Or the problem may be well within their field, but for some reason does not interest them. Or they may be interested, but cannot somehow get to grips with the problem. In all these cases the 'openness' of the problem is a relationship between the problem, the solver, and the occasion, and even the place. It is not any feature of the problem in itself.

> Which of the following grids of squares can be covered by a 'knight's tour'?

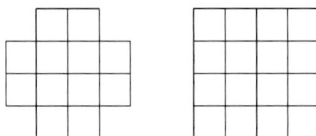

This is so even of those problems that have so far proved more or less 'closed' to all mathematicians. Following the quote from Dieudonné above, he continues by classifying problems according to the kind of success that mathematicians have had with them. His second class is, "Problems without issue (this class includes many problems arising from combinatorics)." Indeed it does, and it once included the problem of the Bridges of Königsberg and many other puzzles which seemed to be unrelated at the time to the mainstream of mathematics, or to any other problem, though that turned out to be a historical and contingent fact, not a necessary one.

Pouring a given quantity from two jugs of known capacities is a well-known problem which will be investigated – I am using the verb, not the noun – by pupils according to their interest, insight of the moment, success, mathematical sophistication … In itself it is not essentially open or closed.

Once again, I suggest, the logical approach, if children are to do mathematics, is to allow them to learn (it is a rather simple lesson) from their own experience of solving problems that some problems attract them more, some are more difficult, that sometimes they are attracted to delve more deeply into a problem, that sometimes they will come to a dead end – which could be the solution which suggested nothing further to them – while at the same time learning via their teacher that these various reactions on their part are perfectly normal and match exactly those of the professionals.

Assumptions, rules and patterns

The misleading characterisation of some problems as 'open', and the consequent effort to search out problems which *are open* or which can be *made open* by the addition of a suitable instruction, has itself had a distorting effect. [Consider] this quote which adds another qualification: "Investigations are those problems in which the procedures for solution are not clearly defined *and where the assumptions made can lead to a variety of outcomes*. The creative processes involved in investigations model closely the methods of the real mathematicians." (5; my emphasis). Is it justified? I suggest it is not, though such an emphasis is becoming very common: "It is also necessary to confront [pupils] with problems where progress towards a solution is not possible until definitions and 'rules of the game' have been discussed and agreed upon" (6).

$$13 = 2^2 + 3^2; \quad 42 = 1^2 + 4^2 + 5^2.$$

Which numbers can be written as a sum of square numbers?

The same theme appears in the *SMILE Handbook* (7): "It is important ... that they have explored a situation governed by certain rules (which they have determined themselves); that they have collected the results together and classified them; and that they have made hypotheses based on these results and then tested them against further results."

There are indeed many professional problems in which assumptions can or must be made leading to different outcomes. But this is not a characteristic of all professional problems. The Riemann Hypothesis is often considered to be the most significant problem in mathematics today. It is remarkably clear in statement, and requires no choice of assumptions.

By adding further qualifications to the idea of an 'investigation', these more recent quotes are becoming more and more one-sided in their picture of what professionals do.

The next statement also from the *SMILE Handbook* adds another qualification, referring to a different kind of rule, in the outcome: "These (investigation) cards generally present a mathematical problem in an open-ended way, leaving the pupil important decisions about direction and depth. There is very little guidance and the children must be encouraged to explore any avenues which they think might lead to patterns or rules ..." Exactly the same comments apply. Patterns and rules are indeed exceptionally important in mathematics, but they are not characteristic of all problem solving. Not all mathematics is inductive. If investigations are meant to be a model of professionals at work, then an over-emphasis on patterns and rules distorts the model. And indeed, this distortion may be observed in the best-known collections of investigations and even more strongly in the written investigations of pupils (8). A very high proportion involve tables of results, (not least because adding an instruction, such as *Investigate other problems like this*, often tends to lead to lists of results and their classification). Fine, but what about other aspects of problem solving which are squeezed out by this distorted presentation? Not only aspects of problem solving but whole areas of content have been left out. It is a notable fact that very few indeed of the investigations in the best known sources relate to the usual syllabus topics.

> On the 5 × 5 square grid, two squares have been shaded. Can the remainder of the grid be covered by 2 × 1 rectangular tiles?

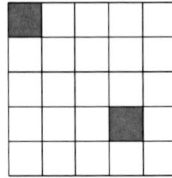

That is unfortunate, I suggest, not because the syllabus is so important in itself, but because much of the richest mathematics has found its way there. That is no coincidence – it was developed by so many mathematicians precisely because it was found to be rich in possibilities. There are wonderful opportunities for pupils' problem solving in standard algebra, for example. And some of these can and will lead to collecting and tabulating results. All of them, without exception are as open as children choose, and are able, to make them. What a tragedy that they are so seldom investigated!

Collection and classification of data is relatively more important in the sciences, hard and soft. Experiment is one aspect only of mathematics. An excellent example of the use of the term 'investigation' based on its scientific use, rather than on an emphasis on openness, making assumptions or deciding on rules, and which does not nevertheless overemphasise pattern-spotting, is the account in Thomas Eastley College's *Mathematical Investigations* (9). The investigations include vibrating strings, a mathematical investigation into the effects of the Black Death on the small Derbyshire village of Eyam during the years 1665 and 1666, a mathematical look at Quorn [in Leicestershire, after which the meat substitute is named] and its surroundings, fractals, music of the bells, permutations and polyhedra.

Learning not to be anxious

The institutionalisation of the term 'investigation' has a further harmful result. The 'openness' of investigations and the emphasis on the need to make assumptions, make decisions, find your own way, all place a strain on many children's senses of autonomy, and easily lead to psychological stress. When pupils solve problems, I am speaking now from my own experience, they will question assumptions, make their own assumptions, take problems which interest them further than expected or in unexpected directions. However, they will initially only do this sometimes, and as a result of their own desires, not because they have already understood the rather difficult notions that are second nature to the professional and which can be acted on, if necessary, in a workmanlike way even when enthusiasm is lacking.

The front page of the excellent magazine *Investigator* published by the SMILE Centre quotes the kind of comments children often make when they are feeling insecure: "More and more teachers are wanting to start using investigations. But there are problems. Children say, What's the point?, What should I do?, What's the answer?, This is silly!, I want to do proper maths!" (10) ... There is a profound irony here. When pupils are told to investigate, they are being told to be free, to behave in a free and unconstrained way, mathematically speaking. What a paradox! Instructing someone to be free (although without sufficient previous experience they may find this freedom difficult to handle) while ignoring their own desires!

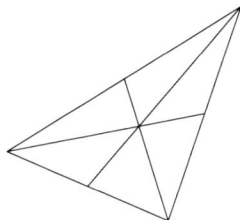

What relationships, if any, are there between the lengths of the lines in this figure?

The 1969 *Bulletin* article (1) recognised the importance of the pupil's own feelings: "There are dangers in presuming that because it is possible to pose a problem, hence people ought to want to solve it. The work *should include some choice* in setting of problems relating to given phenomena." It should indeed, and it should also include a choice on the part of pupils as to how far they pursue the problem. Even if they 'see' further possibilities, they may not be interested to pursue them.

It is indeed desirable that pupils should eventually be as flexible and as sophisticated as possible in their approach to problems. But to begin with, and let me repeat that many young secondary pupils are now doing investigations, it is surely desirable that they should concentrate on the psychologically less stressful task of solving problems which do not deliberately and intentionally throw them into the deep end, but allow them to start from the shallow end, and go in to whatever depth they are ready for. Their teachers might appreciate the difference also!

Summary
- Any distinction between problems and investigations is misleading and not followed by professional mathematicians;
- 'openness' is a function of the problem in relation to the solver, not of the problem alone;

- an emphasis on 'openness' distorts the very process of doing mathematics which investigations are supposed to reveal, while encouraging further distortions;
- a premature emphasis on the solver's autonomy frequently produces stress and anxiety in the solver;
- pupils should rather learn steadily through their own experience how rich and varied problems can be, how open to exploration and interpretation, as well as discovering the approaches which professional mathematicians find helpful – they should never start by being thrown in at the deep end.

References

1. *Sixth Form Mathematics Bulletin No. 1*, ATM, 1969.
2. *Sixth Form Mathematics Bulletin No. 3*, ATM, 1970.
3. Dieudonné, J. A., *Panorama of Pure Mathematics*, Academic Press, 1982.
4. *Points of Departure 1*, ATM (Fourth print, 1982).
5. Stewart, J., *Investigations*, SPLASH 10110, 1979.
6. HMI, *Curriculum 11-16*, HMSO, 1977.
7. *SMILE Handbook*, SMILE Centre, London.
8. I am thinking of *Points of Departure 1 & 2* published by ATM and *Some Ideas for Investigations* published by the SMILE Centre.
9. Thomas Eastley College, *Mathematical Investigations*, ATM, 1979.
10. *Investigator 1*, SMILE Centre, 1984.

1.2 'The investigating experience: Frogs'
by J Dixon and R Watkinson
Vol. 27, no. 5 (November 1998), pp 42-43

The situation

We have seven positions in a row; three on the left occupied by red counters (say) and three on the right by yellow counters; the centre position is vacant.

The object is to change over the counters so that RRR _ YYY becomes YYY _ RRR.

The rules are

1. Red counters move only to the right and yellow counters only to the left.
2. A counter may move into the space next to it if it is empty or may jump over one counter of the other colour into an empty space.

N.B. You do not have to move the colours alternately, and you can start with either colour.

Introduction

We have found that an excellent way to introduce this is to have seven chairs out front, and three boys and three girls exchanging places. This is like a party game. When the pattern of moves has been found and repeated once or twice the children try it individually using counters or pegs and pegboard. Some children will have difficulties and we do not want frustration to set in. To avoid this we can repeat the demonstration. We can let them try with only two a side first and we can bring out in discussion the following points.

1. Sometimes there is only one move you can make; no need to think about it, make it.
2. Sometimes there are two possible moves; then it is important to decide which.
3. Sometimes there are no possible moves; you are blocked. Then you must have gone wrong earlier when you had a choice; try the other alternative.
4. If reds are going to the right and yellows to the left, then RRYY is always a block.

It is important (a) that they don't give up and (b) that they feel they have discovered the solution themselves (however much you might have pushed them in the right direction).

The object is not just to do it once but to know how to do it every time. This points to the need to keep a record of the moves in some way, i.e. they need to invent a code. Some of the codes will be quite complicated. Actually it is enough to indicate the colour of each move, as one never has a choice of move between two of the same colour. So, for example, the solution can be written as RYYRRRYYYRRRYYR with $1 + 2 + 3 + 3 + 3 + 2 + 1 = 15$ moves.

Suppose we now try the puzzle with four-a-side, using the three-a-side solution as a guide. This leads to the solution RYYRRRYYYYRRRRYYYYRRRYYR with number of moves $1 + 2 + 3 + 4 + 4 + 4 + 3 + 2 + 1 = 24$. Can we now predict the solution for the five-a-side puzzle? What about two-a-side? One-a-side?

Now we want to tabulate our results.

No. of counters on left	No. of counters on right	No. of moves needed
1	1	3
2	2	8
3	3	15
4	4	24
5	5	35

Can we predict how many moves the six-a-side puzzle will need? We may see that the differences in the third column follow the sequence 5, 7, 9, 11 and so get the next value as $35 + 13 = 48$. Alternatively, and better, we may go via $15 = 3 \times 5$, $35 = 5 \times 7$ to see that also $24 = 4 \times 6$, $8 = 2 \times 4$ and $3 = 1 \times 3$ so that if we increase the number in column (ii) by 2 and multiply by the number in column (i) we get the number in column (iii); i.e. no. of moves for n-a-side is $n(n + 2)$.

This is perhaps only possible for some pupils. It is possible that someone may recognize the numbers 3, 8, 15, 24, 35, as each being one less than a square number, leading to no. of moves for n-a-side is $(n + 1)^2 - 1$.

Extension

a) Some children might like to investigate what happens when the numbers of the two colours are different.

b) Frogs can now be played and explored further online at the NRICH website, **https://nrich.maths.org/1246** where it is described as a problem for those in the age range 11-14. (*Eds.*)

1.3 Paul Andrews on 'Investigating the structure of Frogs'
Vol. 29, no. 2 (March 2000), pp 7-9

I was fascinated to see the account by Dixon and Watkinson on their work with the Frogs investigation. I suspect that many readers, like me, will be familiar with Frogs as the investigation which informed their experiences of mathematics in the post-Cockcroft era. Indeed, I remember well a local authority advisor basing his in-service courses on the same investigation. These days Frogs has become somewhat neglected and I suspect there may be several reasons for this. Teachers may feel that they have exhausted its possibilities, they may have found other investigations which are fresh to them, they may just be doing fewer investigations with their students. I confess that until recently I had dismissed Frogs as an experience of my younger days and had long since abandoned working with it.

All this changed when a colleague suggested we use it as an introductory activity with a group of first year undergraduate primary teacher trainees. Several of us protested that the students, having been through GCSE, would recognize it but its advocate reminded us that we hadn't seen its use for some years. Moreover, it was an ideal activity for a first session because of its interactivity and overt social implications. Thus we agreed with the proviso that we were to avoid the use of number patterns and focus solely on its underlying structure.

What emerged reversed my perspective on Frogs. For the first time I became aware of its potential for focusing on mathematical ideas like structure, generality, justification and proof rather than number patterns. For the first time I saw why it had been proposed all those years ago. What follows is a brief summary of my students' work which I believe is transferable to the secondary classroom.

Like Dixon and Watkinson I placed seven chairs at the front of the room and asked for volunteers – we used the colours of their tops to distinguish between the two equal groups. Thus we had a team with blue shirts and jumpers playing with a team of greys. I explained the purpose of the task, that the members of each team were to swap places with the members of the other. The rules emerged quite naturally without my having to intervene – a player could *slide* into an empty adjacent seat or *jump* over an adjacent player into the empty seat.

Over the next few minutes several attempts were made to get the two teams past each other. Volunteers offered a range of suggestions and gradually a sense of economy and efficiency emerged – students wanted to complete the swap in as few moves as possible. Thus they realized that jumping over a member of one's own team was inefficient as was moving backwards. Importantly, they noticed that allowing a move which brought two team members together would ultimately mean that one of them would have to move backwards in order to create the necessary space for an opposing team member to jump. This can be seen in the second row of Figure 1. One can see from the first position that either person A or person B can move in their respective forward directions. If A moves then it creates a block which prevents person B from moving forwards. In fact, unless A retreats, the only moves that can be made are C jumping back over A, or D sliding into the space, neither of which are likely to forward the cause for economy.

Gradually a sense of rhythm emerged and trainees began to talk in terms of creating the space for whole groups of players to jump one after the other. Eventually an agreement was reached that the task couldn't be completed in fewer than fifteen moves although there was no sense yet as to why. We added two more chairs, replaced the team members with new ones and continued with teams of four, one in dark and the other in light tops.

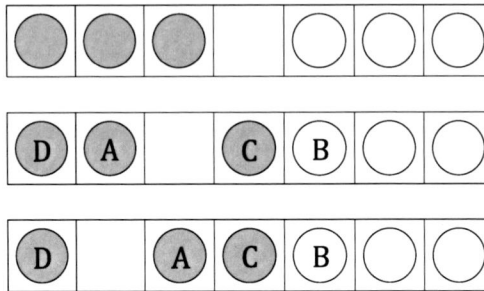

Fig. 1

At one point someone asked if they might record on the board the results they'd got so far. I clarified what the questioner meant and then explained that I wanted them to avoid the use of number patterns and to look for a way of making predictions through an analysis of what they saw their colleagues doing. This caused a few anxious looks but no more.

We continued with four on each side and after a few minutes agreed that 24 was the minimum. Shortly after we'd moved on to five a side – which makes it sound like a game of indoor football – someone announced that she could predict the number of jumps needed for any situation. The offering was delightful in the sense that it was made very tentatively but with a clear sense of excitement and achievement. From any perspective as a teacher this was the defining moment of the lesson. It was the opportunity to make transparent the sense of generality implicit within the particular activity through focusing students' attention on the structural characteristics of their task. I asked the student to explain what she meant. She said that every person in one team had to get past every person in the other team. (In this case) that meant that each individual had to pass four other people and the only way she could do that was by jumping – either they had to jump over her or she had to jump over them. She added that the same was true for every member of the team which meant that sixteen jumps were necessary for two teams of four.

Several people claimed not to understand so I asked someone else to put into their own words what they'd just heard. A second student repeated more or less what had been said but with a few slight shifts of emphasis. A third volunteer came to the front of the room (where eight people still sat) and showed how each person in each team must physically pass all those of the other and that this meant that the number of jumps had to be a square number. Within two or three minutes murmurs of assent were heard around the room. At this point someone asked about the number of slides. A brief discussion led to the conclusion, based on the numerical data collected thus far, that the number of slides might be equal to the total number of players or twice the number in a team.

For colleagues unfamiliar with Frogs a justification for the number of slides is slightly less transparent than for the number of jumps. So I decided to intervene – it was one of those decisions we all make every day. I just felt it would be better for them not to spend time, particularly in their fragile novice state as investigators, on a potentially unrewarding task. So I changed direction and asked if anyone could tell me how many spaces each person would have to move in total. The reply was immediate. Everyone would have to move five spaces because each had to pass the four members of the opposing team plus the empty seat. Thus each team would need to move twenty spaces making a total of forty. I asked how many spaces would be accounted for by the jumps. Several students responded that 32 would be necessary as there were sixteen jumps needed and each accounted for two spaces.

At this point the conversation shifted to one about generality. I asked what they thought I might mean when I talked about the general case. Some students appeared a little confused but several began talking about rules which applied to all cases and which were expressed in terms of n. I asked how we might use our conversations to work on this. The student who had spoken earlier about square numbers suggested that the number of jumps necessary for two teams of n players would be n^2 because each of the n players in one team had to pass each of the n players in the other. Someone else, with a little prompting from me, then pointed out that each player would have to move $n + 1$ spaces and that with $2n$ players in total, n on each side, $2n(n + 1)$ spaces had to be accounted for. I asked how many spaces would be explained by the jumps and was promptly told $2n^2$ because each of the n^2 jumps accounted for two spaces. At this point someone suggested that the number of slides would be the total number of spaces needed less those accounted for by jumps. That is, the number of slides was equal to

$$2n(n + 1) - 2n^2 = 2n^2 + 2n - 2n^2 = 2n$$

which accorded with the conjecture made earlier.

We then spent a few minutes discussing and clarifying what we'd found.

1. The total number of spaces moved by all players is $2n(n + 1)$.
2. The total number of jumps needed is n^2 which accounts for $2n^2$ spaces.
3. The total number of slides is $2n(n + 1) - 2n^2 = 2n$.
4. The total number of moves made by all participants is jumps plus slides which equals $n^2 + 2n = n(n + 2)$.

At this point of the lesson someone made an interesting observation. The student concerned said that when three people in each team were involved different people made different moves. When four people in each team were involved everyone made the same moves. This led to a lengthy discussion on the nature and number of moves made by different people and whether or not they were predictable. So, for example, when there were three people in each team (six in total) the minimum number of moves needed was fifteen which clearly meant that not everyone could make the same number of moves. Someone pointed out that the first person to move had to slide into the vacant seat which meant that he or she, bearing in mind the need to move four spaces in total, had to undergo an even number of slides. I asked the group what this might mean and after a few moments someone made the following suggestion. She said that no player can perform consecutive slides because to do so would lead to two of the same colour being together which, as was discussed above, leads to one of the two having to move backwards. She drew the picture shown in Figure 2 to illustrate what she meant.

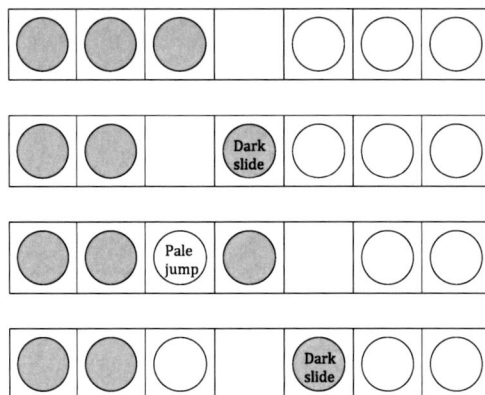

Fig. 2

She continued by saying that if a player cannot make two consecutive slides then the only options in this instance would be for the first player to do a slide-jump-slide routine. She finished by saying that in the case of three players a side only two sequences of moves are possible, slide-jump-slide or jump-jump, and that these occur alternately. A second student offered an alternative perspective. She commented that the efficient solution involved players' initial moves alternating between slide and jump. That is, the first player starts with a slide, the second a jump, the third a slide and so on until all have moved. With a total of six players three must begin with a slide. Therefore, if players must do an even number of slides, the only options are for them to do the same slide-jump-slide or jump-jump routines.

Various analyses of the particular moves made by the members of team of four, five and six followed. These involved students in well-argued discussions in which they were forced by their peers to articulate their reasoning. Thus, for

14

example, one student argued that when each team comprised four players each team member had to move five spaces. Five spaces necessitated everyone doing an odd number of slides. But there were only eight slides available so that everyone would have to do one slide and two jumps. The session ended with small groups discussing what would happen if the numbers of players in each team were different. They were asked not to simulate the game but to analyse the structure in order to arrive at a generality.

Within a few minutes they suggested that with m players on one team and n on the other the total number of moves needed is $mn + m + n$. Importantly, each group of students had arrived at the same solution by invoking the task's structure. They determined the number of jumps needed and the spaces accounted by them. They looked at the total number of spaces needed for the task to be completed and then deducted the number of slides necessary.

Working on Frogs in this way was remarkably empowering. The constant invocation of structure to derive and justify general statements offered insights which would have remained hidden if we'd stuck to number patterns. Importantly, such an approach reminds one that because there is no reason to begin with a simple starting point – one may as well start with a situation representative of the highest, rather than lowest, level of generality. Maybe next year I'll start with different numbers of players in each team.

There were social and pedagogic bonuses which I hoped didn't go unnoticed – students had spent a couple of hours actively, collaboratively and emotionally engaged in a worthwhile piece of mathematics which challenged many of their preconceptions about the nature of mathematics and how it might be constructed. Importantly, not once had anyone put pen to paper nor a number pattern been recorded.

1.4 'An investigation into modulo arithmetic', by Malcolm Swan (written whilst a student at Nottingham University)

Vol. 5, no. 4 (September 1976), pp 30-32

This article brings together several branches of school mathematics while presenting a fresh approach to the subject of modulo arithmetic. In particular, during the course of the investigation, graphs, transformation geometry and group theory all appear. I also hope there is something here to suit every level of ability from the first year of secondary school to the sixth form.

With a younger class I would begin by considering the following 'three times table'.

$3 \times 1 = 3$	$3 \times 6 = 8$
$3 \times 2 = 6$	$3 \times 7 = 1$
$3 \times 3 = 9$	$3 \times 8 = 4$
$3 \times 4 = 2$	$3 \times 9 = 7$
$3 \times 5 = 5$	$3 \times 0 = 0$

The rule that I am here following is "Write down the last digit of every number you come across". So, for example, instead of writing $3 \times 12 = 36$, we simply write $3 \times 2 = 6$. It can easily be shown that subject to our above rule, our table contains every possible entry, for example $3 \times 123 = 369$ appears as $3 \times 3 = 9$. Look at the answers. Every digit from 0 to 9 appears as an answer exactly once. This could lead to a classroom investigation of the various multiplication tables to see which tables do or do not have this property. A natural extension from this simple introduction of modulo 10 arithmetic would be to consider graphical representations of the tables, explaining, for example, how to draw the graph of $y = 3x$ (mod 10). The pattern in this graph is very striking.

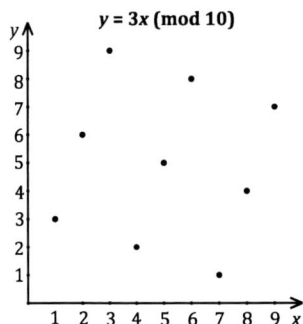

$y = 3x$ (mod 10)

I have omitted the point $x = 0$ because in nearly all the graphs that I will consider, it stays at the origin and is of no interest.) I would ask the class to draw the graphs for the other tables (mod 10). Hopefully, they would notice that the graphs pair off nicely with the notable exception of $y = 5x$ (mod 10). Some children would then possibly see that $y = Ax$ (mod 10) and $y = Bx$ (mod 10) pair off as reflections of one another whenever $A + B = 10$. The reflections are about two possible axes $x = 5$ or $y = 5$. (The latter if we ignore the points lying on the *x*-axis. These remain on the *x*-axis under reflection in the line $y = 5$ because $(x, 10)$ is again on the *x*-axis!)

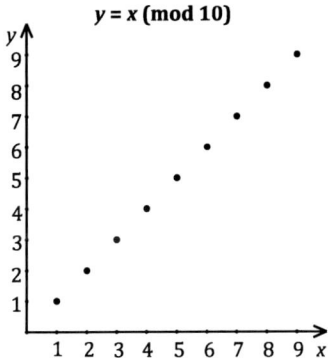

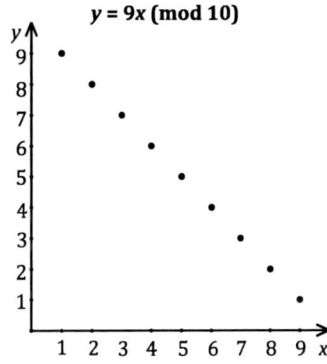

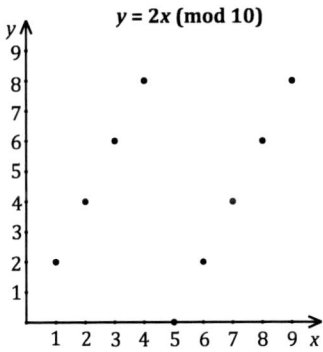

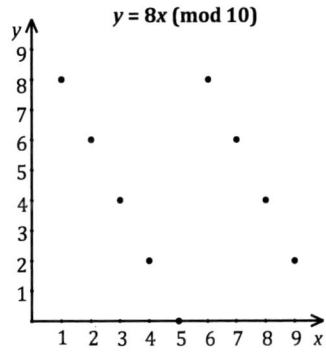

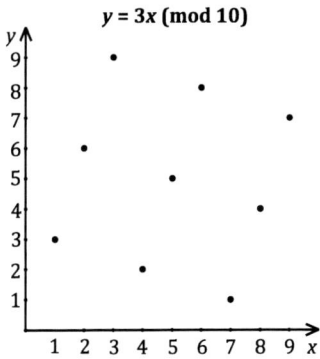

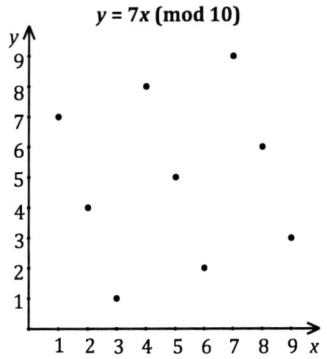

y = x (mod 10)

y = 9x (mod 10)

y = 2x (mod 10)

y = 8x (mod 10)

y = 3x (mod 10)

y = 7x (mod 10)

17

y = 4x (mod 10)

y = 6x (mod 10)

y = 5x (mod 10)

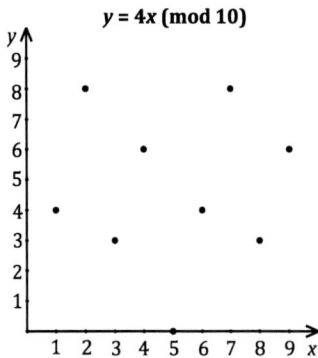

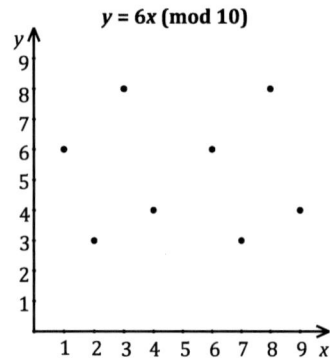

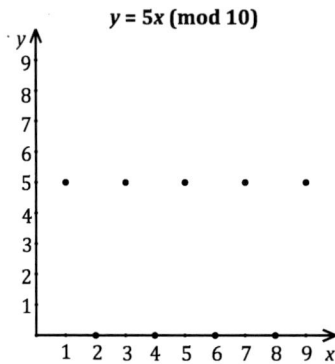

I will digress to prove these two results:

The pairs are reflections of one another in the line $x = 5$

Proof: $Ax = 100 - 10A - 10x + Ax \pmod{10}$
$\Leftrightarrow Ax = (10 - A)(10 - x) \pmod{10}$
$\Leftrightarrow Ax = B(10 - x)$ (since $A + B = 10$).

So the point (x, Ax) on one graph corresponds to the point $(10 - x, B(10 - x))$ on the other graph giving the line of reflection $x = 5$.

The pairs are reflections of one another in the line $y = 5$

Proof: $(A + B)x = 0 \pmod{10}$ (since $A + B = 10$)
$\Leftrightarrow Ax + Bx = 0 \pmod{10}$
$\Leftrightarrow Ax + Bx + (10 - Bx) = (10 - Bx) \pmod{10}$
$\Leftrightarrow Ax = (10 - Bx) \pmod{10}$.

So the point (x, Ax) on one graph corresponds to the point (x, Bx) on the other by a reflection in the line $y = 5$.

If instead of just considering modulo 10, we looked at, say, modulo 7, all the results discovered above would remain valid. (There would still be two axes of reflection, i.e. $x = 3\frac{1}{2}$ and $y = 3\frac{1}{2}$). This would be the next step in the investigation and would hopefully give the children some practice at working with different moduli. Another extension that could be followed is to look at a pair of graphs, say $y = 3x$ (mod 10) and $y = 7x$ (mod 10) and to see what happens as we increase the power of x. The pair of graphs $y = 3x^2$ (mod 10) and $y = 7x^2$ (mod 10) have lost the reflection property in $x = 5$ but have retained the line of reflection $y = 5$. Both lines appear in $y = 3x^3$ (mod 10) and $y = 7x^3$ (mod 10) but the line $x = 5$ disappears again in the quartics. A very surprising thing occurs when we reach the pair $y = 3x^5$ (mod 10) and $y = 7x^5$ (mod 10). Suddenly we find ourselves with the same graphs as $y = 3x$ (mod 10) and $y = 7x$ (mod 10), but more of that later!

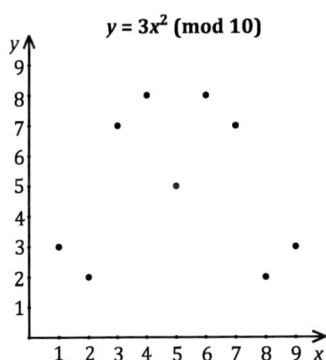

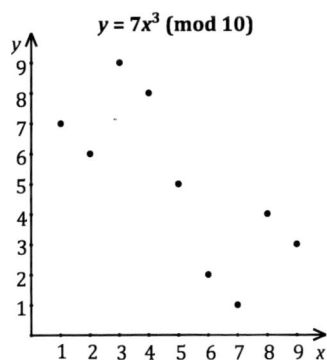

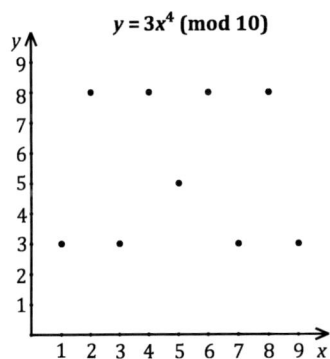

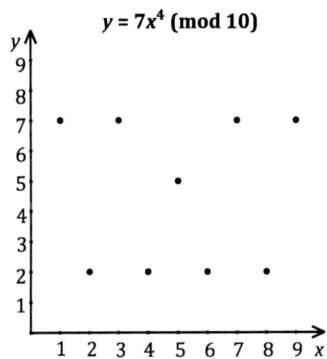

$y = 3x^2$ **(mod 10)**

$y = 7x^3$ **(mod 10)**

$y = 3x^4$ **(mod 10)**

$y = 7x^4$ **(mod 10)**

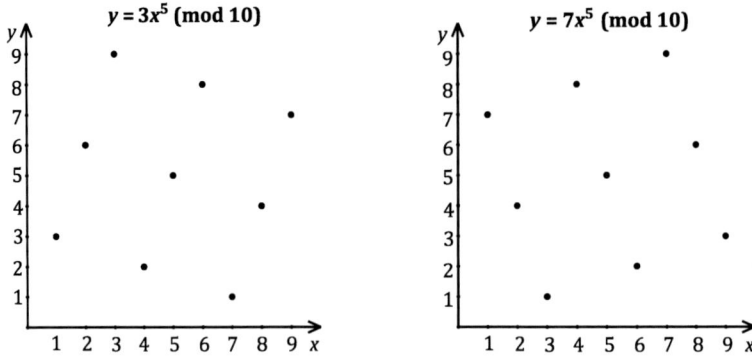

$y = 3x^5 \pmod{10}$ $y = 7x^5 \pmod{10}$

In general, $y = Ax^i \pmod{m}$ and $y = Bx^i \pmod{m}$ are related when $A + B = m$ by two reflections if i is odd and by just one if i is even. (The proof involves a similar procedure to that done in the case $i = 1$, $m = 10$ and is not difficult.)

This leads nicely on to the next part of our investigation. I decided to look at simple polynomials to see what patterns appear and how they change as the index of x increases. This line of research does in fact lead to some very surprising results. In the classroom, I would carefully show the children how to avoid evaluating things like $8^8 \pmod{9}$ by laboriously working out 8^8 and successively dividing by 9s to find the remainder! These calculations are unnecessary if each polynomial graph is constructed from the previous one. Let me illustrate this with an example. Suppose we wish to plot the graph $y = x^5$ $\pmod{7}$. To do this we use the graph of $y = x^4 \pmod{7}$; see Fig. 3. When $x = 1$, $y = 1$ for all powers, so there is no problem! Where $x = 2$, we use the $y = x^4$ $\pmod{7}$ graph to find $x^4 = 2 \pmod{7}$.

So $x^5 = x \times x^4 = 2 \times 2 = 4 \pmod{7}$ and so we plot the point $(2, 4)$. After a little practice, this becomes easy to do in one's head! On the next two pages, I have drawn the graphs of polynomials for the first few moduli in this way.

The following interesting results emerge:

Property 1 Whatever the modulus, all the graphs of even powers of x are symmetrical about the mid-point of the x-axis (i.e. about the line $x = \frac{m}{2}$, where m is the modulus).

Property 2 The patterns repeat after a certain power of x. (For example, take modulus 6, we get graphs of x^0, x^1, x^2 all different and then the graph of x^3 is identical to x^1, x^4 is identical to x^2, x^5 is identical to x^1 and so on. These two patterns will repeat forever as will be seen when we consider that each graph can be drawn from the previous one.)

Property 3 If a point on the $y = x^r$ graph lies on the x-axis, it will remain there on the $y = x^s$ graph where $s > r$ (for a given modulus). Such a point, I will call dead.

Modulo 4

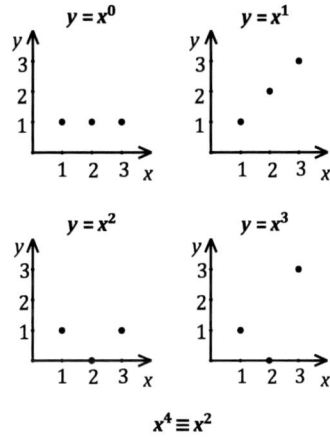

Modulo 3

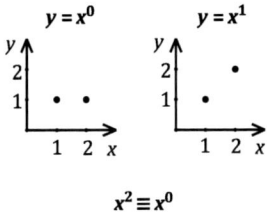

$$x^2 \equiv x^0$$

$$x^4 \equiv x^2$$

Modulo 5

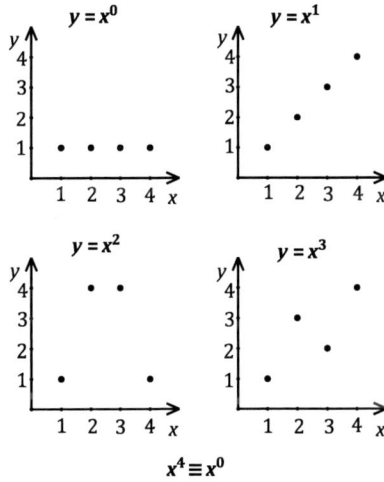

$$x^4 \equiv x^0$$

Modulo 6

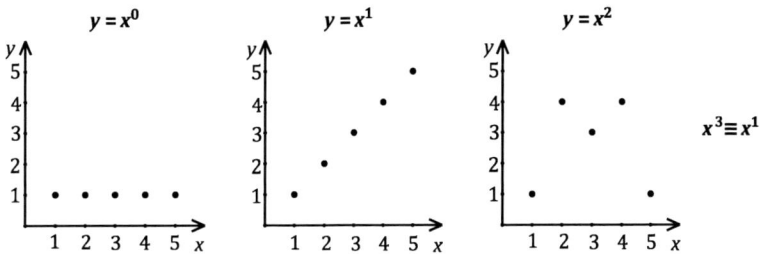

$$x^3 \equiv x^1$$

21

Modulo 7

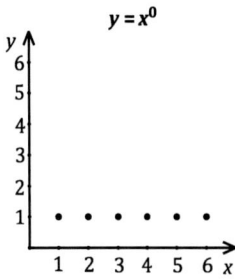

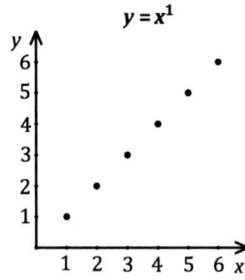

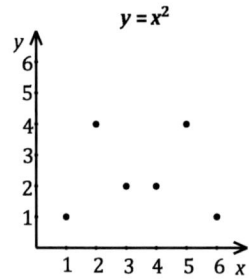

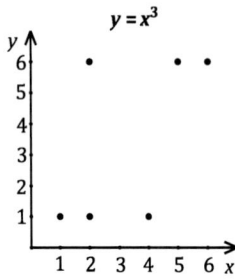

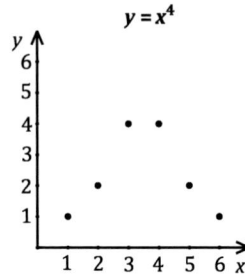

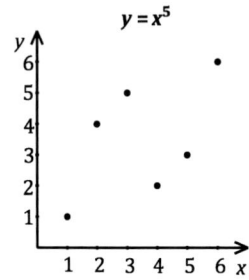

$$x^6 \equiv x^0$$

Modulo 8

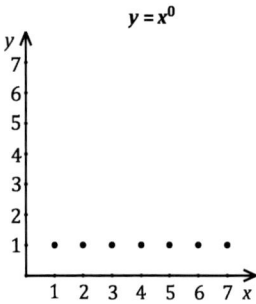

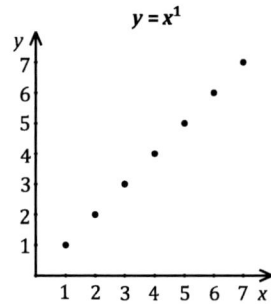

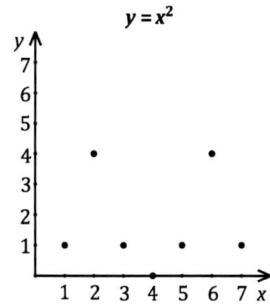

22

$y = x^3$ $y = x^4$

$x^5 \equiv x^3$

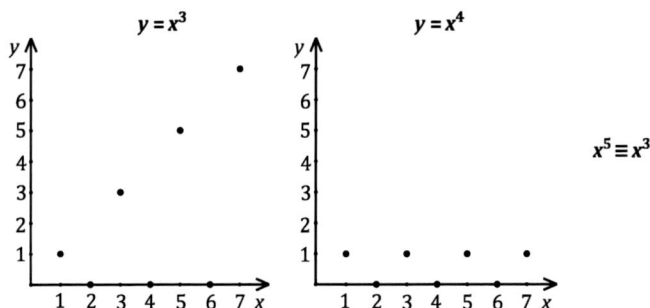

Property 4 Every pattern has some form of rotational or reflectional symmetry if we ignore dead points. This seems remarkable! These properties are all fairly obvious from drawing the graphs, but are not at all obvious when it comes to prove them. I have drawn up a table showing the number of distinct graphs that appear for each value of the modulus. Can you see any pattern at all?

Modulus	2	3	4	5	6	7	8	9	10	11	12	13
No. of distinct graphs	1	2	4	4	3	6	5	8	5	10	4	12

I have been unsuccessful, so far, in trying to find a rule, but one or two things can be pointed out. All those values of the modulus which are prime (say p) have exactly $p - 1$ distinct graphs. We may ask whether this an accident? Looking again at the polynomial graphs, it is noticeable that when we are dealing with a prime modulus p, then the graph of $y = x^{p-1}$ (mod p) is identical to the graph of $y = x^0$ (mod p), or to put it another way

$$\forall x, x^{p-1} = x^0 \text{ (mod } p).$$

This result is in fact a statement of Fermat's famous number theory theorem. (Leibniz gave the first known proof.) However, this does not prove that there are exactly $p - 1$ distinct graphs, it just shows that there are at most $p - 1$ distinct graphs. Euler generalised Fermat's theorem by considering any modulus m, then $x^{\phi(m)} = 1$ (mod m) provided x and m have no factors in common ($\phi(m)$ here means the number of integers which have no common factors with m.) This gives another upper bound for the number of distinct graphs.

1.5 'The investigating experience: Consecutive sums' by J Dixon and R Watkinson

Vol. 26, no. 5 (November 1997), pp 13-15

The task

To find out which numbers can be written as the sum of two or more consecutive integers.

Introduction

This activity is suitable for a wide range of age and ability and can be carried out with the whole class or a small group if preferred.

We first need to establish the idea of 'consecutive', using examples such as days, months, letters and numbers. Then we ask the children if they can find two consecutive numbers which add together to make 5, say. We write $5 = 2 + 3$ and this itself may promote discussion because many children read '=' as 'makes'. Can we find some more numbers like 5, which can be written like this? e.g. $13 = 6 + 7$, etc. Can all numbers be expressed as the sum of two consecutive integers?

We need to record our results systematically and one way is to write the numbers from 1 to 20, say, in a column, thus:

```
1
2
3 = 1 + 2
4
5 = 2 + 3
6
7 = 3 + 4
...
```

It soon becomes clear that you cannot make even numbers this way (why not?) and that nearly all the odd numbers can be made. The exception is 1. Can we make 1? What if we use zero? Then the pattern is complete. At this point the teacher might like to ask the children if, given an odd number, say 35, they can find the two consecutive numbers whose sum will make it.

So far we have only considered sums of two consecutive integers and have found that some numbers (i.e. the evens) cannot be made. Can we make them using more than two? This may have already been noticed e.g. $6 = 1 + 2 + 3$ and $10 = 1 + 2 + 3 + 4$. The investigation is now open to the children.

If they work systematically there are several patterns for them to notice, as the table below shows. The starting numbers in each column are the triangular numbers. The interval between successive numbers that can be made goes up by one each time. The numbers which do not appear in any of the columns are the powers of 2 (2, 4, 8, 16, etc).

	Sum of 2 consecutive integers	Sum of 3 consecutive integers	Sum of 4 consecutive integers	Sum of 5 consecutive integers
1	0 + 1			
2				
3	1 + 2	0 + 1 + 2		
4				
5	2 + 3			
6		1 + 2 + 3	0 + 1 + 2 + 3	
7	3 + 4			
8				
9	4 + 5	2 + 3 + 4		
10			1 + 2 + 3 + 4	0 + 1 + 2 + 3 + 4
11	5 + 6			
12		3 + 4 + 5		
13	6 + 7			
14			2 + 3 + 4 + 5	
15	7 + 8	4 + 5 + 6		1 + 2 + 3 + 4 + 5
16				
17	8 + 9			
18		5 + 6 + 7	3 + 4 + 5 + 6	
19	9 + 10			
20				2 + 3 + 4 + 5 + 6

Extension

Given any number, say, 39, can you decide whether and how it can be expressed as the sum of consecutive integers? What about 50? The children could choose some starting numbers of their own and show if and how they can be made as the sum of consecutive integers.

1.6 'Transforming *Billiards* into *Diagonals*' by Dietmar Küchemann
Vol. 14, no. 1 (January 1985), pp 2-3

In the course of introducing student teachers to mathematical investigations, I have become increasingly fond of two particular investigations: *Diagonals* and *Billiards*.

DIAGONALS

On squared paper draw a rectangle five squares by six squares. Draw a diagonal of this rectangle.

How many squares does the diagonal pass through?

Do this for other rectangles.

Can you forecast the number of squares passed through if you know the length and width of the rectangle?

BILLIARDS

This billiard table only has four pockets and the base is divided into squares. Only one ball is used and it is always struck from the same corner at 45° to the sides. (The ball always rebounds at 45° to the sides.)

Can you work out which pocket the ball will fall into?

What would happen if the billiard table was of a different size?

On some tables the ball will travel over every square. Which tables?

How many times does the ball hit the side of the table?

One reason for using them year after year is that these investigations (but not their various solutions!) readily come to mind. I don't know who thought them up, but *Diagonals* is given an extensive treatment in *Starting Points* by Banwell, Saunders and Tahta (published back in 1972 but now sadly out of print); *Billiards* appears as a glossy centrefold in *Factor 5*, a comic-come-magazine produced by the SMILE centre. Both investigations have the useful characteristic of readily engendering rules that break down. For example, in *Diagonals* the number of traversed squares appears to be one less than the sum of the dimensions of the rectangle, as in all but the last case below.

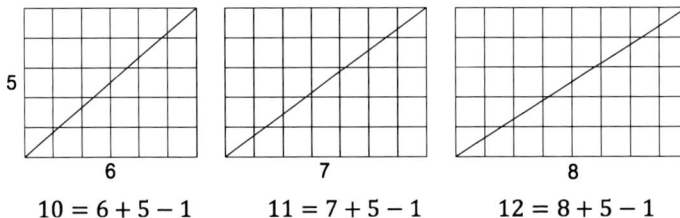

$$10 = 6 + 5 - 1 \qquad 11 = 7 + 5 - 1 \qquad 12 = 8 + 5 - 1$$

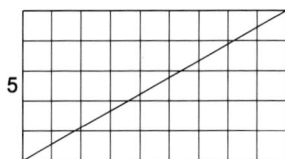

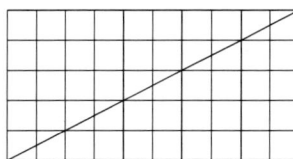

$13 = 9 + 5 - 1$ $10 \neq 10 + 5 - 1$

Looking back through *Starting Points*, I now discover that *Billiards* is in there too (as 'Rebounds'). Indeed, it appears on the same page as *Diagonals*, linked by the common themes of prime numbers (page 33) and square grids (page 209). Clearly, my liking both investigations is no coincidence. To my further delight, a recent re-working of *Billiards* has uncovered a remarkable underlying unity between the two tasks. This can be seen if *Billiards* is 'folded out':

For a 4×3 rectangle, say, the standard diagram for the path looks like this:

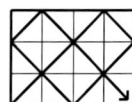

Now consider a representation where the path is built up one stage at a time. (This is given on the next page, not immediately below. *Eds.*)

We end up with Diagram A (below) transformed into Diagram B. In turn, B can be transformed into C by applying a pair of one-way 'shrinks'.

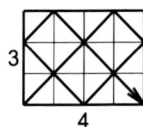

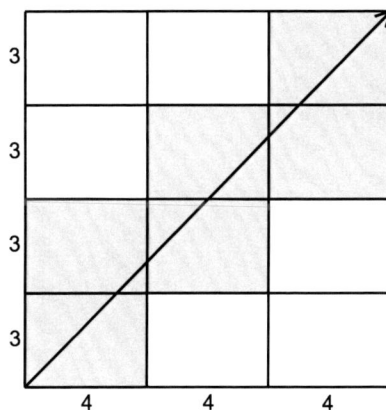

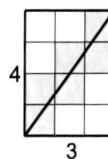

A   B   C

Comparing Diagrams A and C, the question "How many stages on a 4×3 table?" can be seen to correspond precisely to "How many squares does the diagonal pass through on a 3×4 rectangle?" I leave it to the reader to consider how Diagram B, in particular, can be used to answer some of the original Billiards questions; and to investigate the relationship between Diagrams A, B and C when the dimensions of the billiard table are not co-prime.

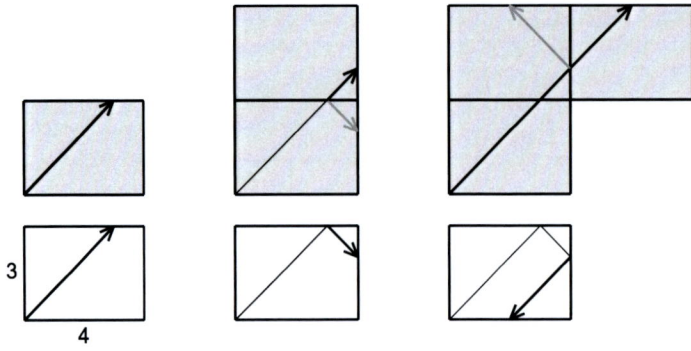

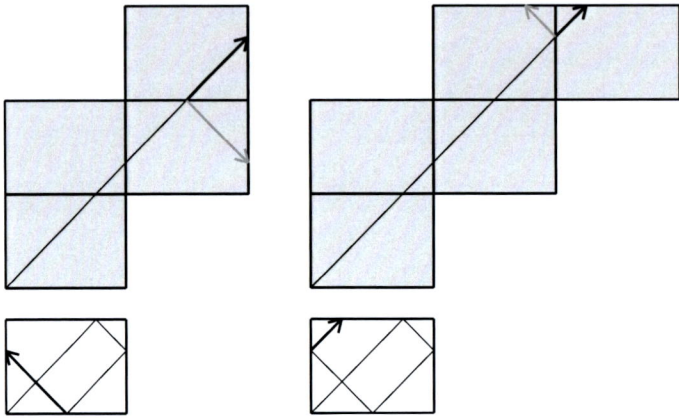

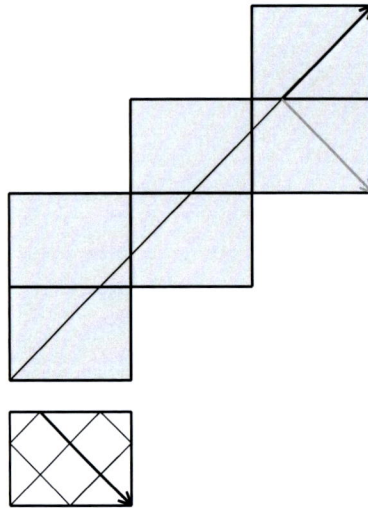

1.7 Doug French on 'Investigating volumes'
Vol. 17, no. 3 (May 1988), pp 27-29

A popular problem for investigation, which was cited in *Mathematics from 5-16* (1), concerns a rectangle of card from which the corners are cut out and the resulting flaps folded up to make an open-topped box.

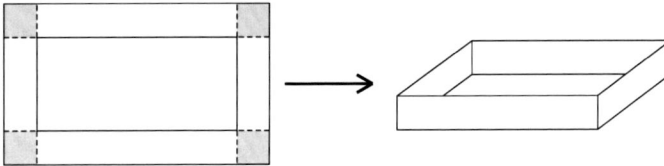

Fig. 1

The usual problem is to determine the size of squares to cut out to give the maximum volume, given the dimensions of the rectangle of card. The problem can be tackled at many levels and there are various ways in which it can be altered or extended. It also offers the opportunity for some practical work, for it is a useful exercise to make models of some of the different possibilities.

I should like to suggest in this article some alternative problems involving boxes of different designs which provide similar opportunities for investigation and practical work at different levels.

If you dismantle most supermarket boxes used for packing tins and other items, the net from which they are made is found to be a rectangle, if the narrow flap on one edge used for stapling or gluing is ignored. The top and bottom of the box are formed from wide flaps along the top and bottom edges of the rectangular net and these overlap to provide added strength and a surface for fixing.

The width of these flaps is half the width of the box, so using *l*, *w* and *h* for the length, width and height of the box, the dimensions of the net are as shown in Figure 2. It is essential when pupils are investigating such a box that they have the opportunity of examining an actual box to see its structure and, in most cases, they will benefit from constructing some model boxes of their own, to an appropriate scale.

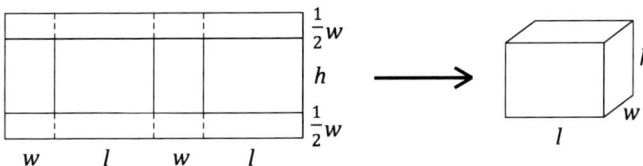

Fig. 2

I give here these possible starting points for investigating this type of box.

Investigation 1

Investigate the volume of a box made from a rectangle of card 100 cm long and 40 cm wide. Most children will start with a particular example and will rapidly find that the height and length can be readily determined from the width. For a width of, say, 20 cm, the length is found to be 30 cm and the height is 20 cm. The volume is therefore $30 \times 20 \times 20 = 12000$ cm^3 An obvious next move is to tabulate results for varying widths and graph the results. It becomes evident in doing this that the width cannot exceed 40 cm, if the height is not to become negative. I give below a table taking values of w from 0 to 40 at intervals of 5 cm, together with a graph displaying the results.

w	l	h	V
0	50	40	0
5	45	35	7875
10	40	30	12000
15	35	25	13125
20	30	20	12000
25	25	15	9375
30	20	10	6000
35	15	5	2625
40	10	0	0

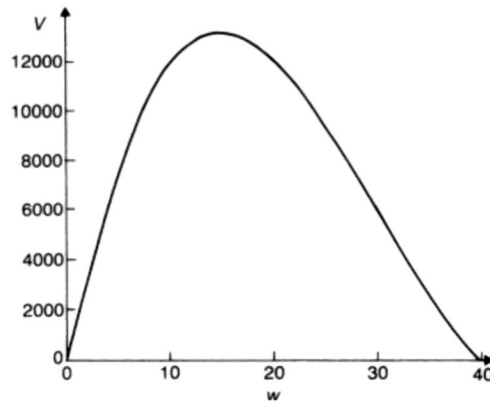

Fig. 3

The table seems to suggest a maximum volume of 13125 cm^3, when the width is 15 cm, and the temptation is to draw a graph with that as a maximum point. However, children should be challenged to investigate further! Further experiments with a calculator will lead to the fact that the maximum volume is given by a value of w slightly below 15. A width of 14.5 cm gives a volume of 13126 cm^3, to the nearest whole number, compared with 13125 cm^3 for a width of 15 cm. Further calculation shows that the maximum volume is given by a value of w between 14.7 and 14.8.

Somebody might suggest doing the calculations on a computer and some children can be encouraged to write a suitable short program (hereafter updated to 'code', *Eds.*). To do this it is necessary to express the relationships algebraically to determine h and l in terms of w. This can be done as follows:

$$w + l + w + l = 100 \implies w + l = 50 \text{ or } l = 50 - w.$$
$$\frac{1}{2}w + h + \frac{1}{2}w = 40 \implies w + h = 40 \text{ or } h = 40 - w.$$

These two results will have been used in calculating the table of values previously, but will probably not have been expressed in an explicit algebraic form. Some code can now be produced. This allows further results to be generated and might prompt further questions, such as what happens when values of w outside the range 0 to 40 are input. This could lead on to a look at the cubic function $V = w(40 - w)(50 - w)$ part of whose graph has been drawn previously (Fig.3).

Investigation 2

The box is to hold 24 baked bean tins of diameter 7.5 cm and height 11.5 cm. Investigate suitable arrangements of the tins and the size of box needed. In this case the volume is effectively determined by the number of tins and the dimensions have to be multiples of the diameter and height of the tins. A typical arrangement would be 4 tins long, 3 tins wide and 2 tins high which gives:

$$l = 4 \times 7.5 = 30,$$
$$w = 3 \times 7.5 = 22.5,$$
$$h = 2 \times 11.5 = 23.$$

The rectangle of card then has

$$\text{length} = 2(30 + 22.5) = 105 \text{ cm},$$

$$\text{width} = 23 + 22.5 = 45.5 \text{ cm}$$

which gives Area $= 105 \times 45.5 = 4777.5 \text{ cm}^2$.

Other arrangements of tins can be considered in the same way and the four results using the least area of card are tabulated below.

No. of tins	Length (cm)	Width (cm)	Area (cm^2)
4 by 3 by 2	105	45.5	4777.5
4 by 2 by 3	90	49.5	4455
6 by 2 by 2	120	38	4560
3 by 2 by 4	75	61	4575

As before, some children can be encouraged to write some code which gives the area when a particular arrangement is input or, better, which enumerates all the possibilities together with areas of card required.

This particular problem can provoke some interesting discussion, because there is not an obvious best arrangement, for a case can be made for several of the possibilities. It is particularly instructive to make scale models of several of the boxes to emphasise that the area of card required is not the only consideration in designing a box. Factors such as stability, strength of the base, ease of carrying and aesthetic considerations all need to be taken into account. It is a valuable exercise to consider all these factors and to arrive at a reasoned conclusion as to which is best, acknowledging that others may quite reasonably come to a different conclusion.

Investigation 3

The box is to have a volume of 20000 cm^3, without the restrictions imposed by having to hold tins. However, the problem is complicated without some further restriction and two possibilities are considered below.

(a) The ends could be square, in which case $h = w$. As in the previous examples the next step for most children will be to look at some numerical examples and then tabulate some results. For a width of 20 cm, the length of the box can be determined by dividing the volume by the area of one end, giving

$$20000 \div 20^2 = 50 \text{ cm}.$$

The dimensions of the rectangle of card can then be calculated:

$$\text{length} = 2(20 + 50) = 140 \text{ cm},$$

$$\text{width} = 20 + 20 = 40 \text{ cm}$$

$$\text{area} = 140 \times 40 = 5600 \text{ cm}^2.$$

A few such results are tabulated below with a rough graph plotted alongside:

w	Area (A)
10 cm	8400 cm^2
20 cm	5600 cm^2
30 cm	6267 cm^2
40 cm	8400 cm^2

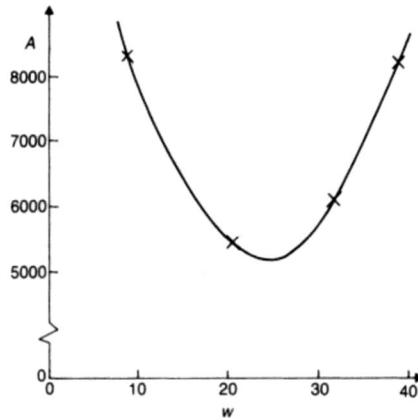

Fig. 4

The general shape of the graph suggests that the minimum area arises for a value of w between 20 and 30, and that it is nearer to 20 than 30. Further calculation will lead to a more precise result.

Obviously, the problem can be approached algebraically using the calculus, but if it is considered numerically it becomes accessible to a wider range of children. Some children can approach an algebraic formulation through trying to write some suitable code to calculate results.

(b) Another variation is to let the length and width be in a fixed ratio. The case of a square base, $l = w$, gives precisely the same results as a square end and other cases require a similar approach so they are not developed in detail here.

The last two investigations concern the traditional net that is used for making a cuboid. This can be cut conveniently from a rectangle of card. Figure 5 shows the arrangement and indicates how the dimensions of the card are related to the dimensions of the box.

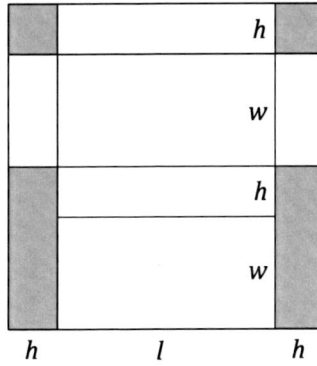

Fig. 5

Investigation 4

Investigate the volume of a cuboid made from a square of card with 20 cm edges. This is very similar to Investigation 1 with $h = 10 - w$ and $l = 2w$, which gives $V = 2w^2(10 - w)$, a cubic function whose graph is sketched in Figure 6.

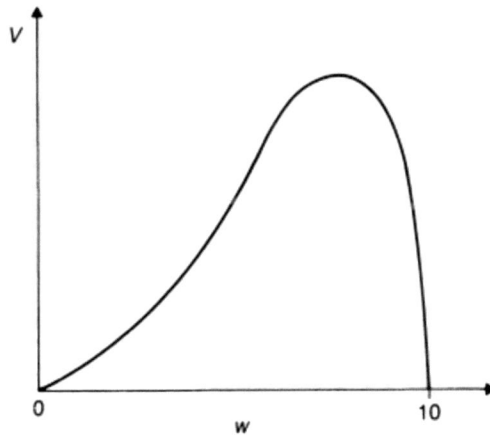

Fig. 6

Children can, of course, investigate the problem numerically as in the previous examples.

Investigation 5

Investigate the area of card required for a volume of 100 cm^3. Here there are similarities to Investigation 3 and the necessity of imposing a further restriction. If $l = 2w$, the card is square and $V = 2w^2h$. The length of the edge is $2(w + h)$, which leads to an expression for the area:

$$A = 4\left(w + \frac{50}{w^2}\right)^2.$$

The graph of this function is similar in shape to that shown in Figure 4 and the value of w to give minimum area can be investigated numerically. It is not suggested that any group of children should consider all these problems, but that they provide alternative variations on a similar theme. They involve an investigative approach, offer plentiful opportunities for fruitful discussion together with the chance to engage in practical work making models to illustrate the various possibilities. A number of essential ideas from the normal mathematics syllabus are involved and these are presented in a meaningful and interesting context.

Reference

1. HMI, *Mathematics from 5 to 16*, DES, 1985.

1.8 Peter Critchley on 'Beautiful magic squares'
Vol. 19, no. 1 (January 1990), pp 48-49; extract

In the preamble, not included here, the author referred to the magic square:

8	1	6
3	5	7
4	9	2

and continued ...

How many teachers, let alone children, have noticed that:

- numbers in the opposite corners (diagonally) add to make 10, twice the centre number.
- the opposite pairs of middle numbers also total 10.
- the four corner numbers total the same as the four middle numbers.
- 4 + 3 = 1 + 6 (notice the position of the numbers)
- 4 + 9 = 7 + 6 (notice the position of the numbers)
- 8 + 3 = 9 + 2 (notice the position of the numbers)
- 7 + 2 = 1 + 8 (notice the position of the numbers)
- 8 − 3 = 7− 2 (notice the position of the numbers)
- 4 - 3 = 7 - 6 (notice the position of the numbers)
- the total of the 8,1,5,3, group is 2 less than the total of the 1,6,5,7, group which in turn is 2 less than the total of the 3,5,9,4, group which is 2 less than the total of the 5,7,2,9, group.

However, what has interested me more than anything else has been the number of patterns that can be found along certain rows, columns or diagonals and I have summarised them on the following diagram.

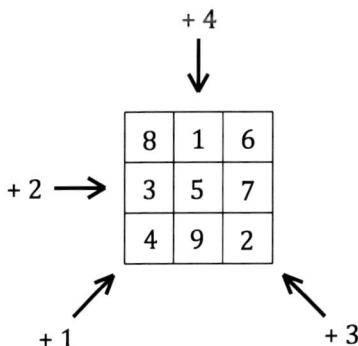

The immediate thought I had was, could I build upon these patterns in order to produce a magic 5 by 5 square? My initial conjecture was that the 1 and 25 and the number half-way between, namely 13, should be placed as shown:

		1		
		13		
		25		

I then decided that as the one particular diagonal on the 3 by 3 went up in 3s, then perhaps the diagonal of the 5 square should go up in 5s, so:

23		1		
	18			
		13		
			8	
		25		3

It seemed obvious at the time that the middle column should increase in steps of 6 and this fitted in 'beautifully' with the numbers already placed.

23		1		
	18	7		
		13		
		19	8	
		25		3

As the middle row of the 3 by 3 went up in 2s, one less than the diagonal, I guessed that in the case of the 5 by 5, the increase should be of the order 4, so:

23		1		
	18	7		
5	9	13	17	21
		19	8	
		25		3

To keep to the pattern suggested by the 3 by 3, I thought the remaining diagonal should increase by 3 each time, so:

23		1		19
	18	7	16	
5	9	13	17	21
	10	19	8	
7		25		3

Unfortunately this system produced two numbers I had already. This proved somewhat of a disappointment and so I applied the technique to a 7 by 7 to see whether duplication was common.

+ 8
↓

+ 6 →

46			1			40
	39		9		35	
		32	17	30		
7	13	19	25	31	37	43
		20	33	18		
	15		41		11	
10			49			4

+ 5   + 7

From the results you can see that it wasn't. I later proved that duplication was unique to the 5 by 5. I felt now that perhaps the diagonal from bottom left to top right should perhaps consist of consecutive numbers and so I started all over again. Here is what I ended up with.

23		1		15
	18	7	14	
5	9	13	17	21
	12	19	8	
7		25		3

46			1			28
	39		9		27	
		32	17	26		
7	13	19	25	31	37	43
		24	33	18		
	23		41		11	
22			49			4

This approach resulted in neither of the squares containing any duplicate numbers but there was no way the remaining unused numbers could be fitted in to make magic squares. Can you see why? However, I suddenly spotted that both the 5 by 5 and the 7 by 7 contained a complete 3 by 3 and what's more they were magic! In both cases too the magic number was three times the centre number as was the case for the original 3 by 3. Looking carefully at the arrangement of the numbers made me think that I had hit on another method for generating 3 by 3 magic squares, the centre number being $\frac{1}{2}(s^2 + 1)$ where s is any odd number. The completed magic square would therefore be of the form:

$\frac{1}{2}(s^2 + 1) + s$	$\frac{1}{2}(s^2 + 1) - (s + 1)$	$\frac{1}{2}(s^2 + 1) + 1$
$\frac{1}{2}(s^2 + 1) - (s - 1)$	$\frac{1}{2}(s^2 + 1)$	$\frac{1}{2}(s^2 + 1) + (s - 1)$
$\frac{1}{2}(s^2 + 1) - 1$	$\frac{1}{2}(s^2 + 1) + (s + 1)$	$\frac{1}{2}(s^2 + 1) - s$

Studying the arrangement even further, I came to the conclusion that it might be possible to generate a magic square by starting with any number in the middle of the top row. The steps for doing so are shown on the next page:

	11	

any number

	11	
13		
		12

next two consecutive numbers placed as shown

+ 9 (

	11	
13		
22		12

add any number to the 13, say 9

	11	24
13	23	
22		12

next two consecutive numbers after 22

	11	24
13	23	33
22		12

) + 9

add 9 here also as you added 9 previously

34	11	24
13	23	33
22	35	12

enter next two consecutive numbers after 33

This suggests that the magic square is of the form:

$x + 2y + 5$	x	$x + y + 4$
$x + 2$	$x + y + 3$	$x + 2y + 4$
$x + y + 2$	$x + 2y + 6$	$x + 1$

where y is the 'any number you add' at stage 3. I now leave it to the reader to compare these magic squares with the original one. What similarities can you find? Do you find these sorts of discoveries beautiful.

2   *Geoboard Investigations*

2.0 Introduction

The mathematics of geometric shapes and the trigonometry associated with them can be quite a dull subject for students. Exercises involving the identification of types of polygons, exploring their symmetry, calculating lengths and areas are difficult to set in some form of context. Yet, the learners have an immediate goal if these aspects are encountered within a mathematical investigation. And there are often several different routes to follow in an investigation leading to excellent opportunities for students to work cooperatively on subtasks within the investigation.

The first three articles in this chapter demonstrate these ideas very well. The theme in each case is a 'polygon hunt'. Students are encouraged to find the shapes of different polygons that can be drawn on a three-by-three grid of points. The first article by William Ewbank shows the number of different polygons that his students found with some simple trigonometry required to calculate the sum of the interior angles. Students find Pick's theorem quite fascinating and this is the theme of the article by James Dunn. The article by Alan Burns develops more advanced mathematical thinking skills using 'The polygon hunt' to explore a proof for the maximum number of n-sided figures and hence of all figures possible on a three-by-three geoboard. Though it confirms many of the findings in the articles by Ewbanks and Dunn, we have placed it at the end of the chapter because of its scope and length.

The grid points do not need to form a square. The article by Nigel Corps and his co-authors uses the points forming a circle and Dave Kirby uses a seven-pin geoboard.

Geoboards were invented by the Egyptian mathematician Caleb Gattegno and became popular in the 1950s as a fun activity for making shapes and considering their properties. In the traditional form they were wooden boards with small nails or pegs placed in rows and shapes could be formed using string or rubber bands. They could often be found in primary schools as mathematical manipulatives for exploring basic shapes. The article by James Bidwell and David Hale shows their use in a primary school setting for exploring areas. The wooden geoboard is now replaced by grids of points on paper and is also available as an app for iPads etc.

The simple idea of a geoboard still provides an excellent means of investigating geometric ideas in a fun and creative way.

2.1 William A. Ewbank on a 'Polygon hunt'
Vol. 3, no. 4 (July 1974), pp 16-17

Are you a doodler? Mathematicians and artists make the best doodlers, as their doodling can be creative. The only trouble with mathematical doodling is that it soon becomes too intellectually absorbing to enable the doodler to concentrate on the speaker of the moment, thereby disqualifying it as doodling within the meaning of the term. An interesting piece of doodling is to take a 3×3 geoboard – no, not literally; a 3×3 array of dots will do – and make all possible closed curves on it. These would be polygons, of course, some concave or re-entrant. So I call it 'Polygon hunt'. I have given this task to older students, and they have come up with 50-60 different polygons in 45 minutes. At one workshop for teachers, I divided the participants into four groups who competed to see which could come up with the most different polygons in a given time. To do this, they had a few geoboards (optional), and sheets of 3×3 dot arrays apiece. Ordinary dot paper would do. One person in each group acted as scribe, and took the valid polygons, and redrew them on a larger 3×3 dot array, which was then stuck on a blackboard according to its type – triangle, quadrilateral, pentagon, hexagon, or heptagon. When "time" was called, each section counted its totals, and everyone went round checking the other groups to see that no duplicates had been included. Like the Dodo's race in Alice, everyone won, and everyone had a prize, the fun of sharpening one's powers of perception and geometrical wits. Indeed, a sharp perception is needed to compare shapes, and check for duplication. A rule was made that shapes that are congruent by rotation or reflection count as one. For example, the shapes in Figure 1 would be equivalent.

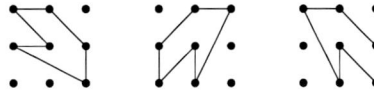

Figure 1

Some students expressed surprise that all the 6-sided shapes were named 'hexagons'. To them, 'hexagon' meant only a regular hexagon. Something learned here. After all possible polygons have been found, it is possible to classify them into many different subsets, such as:

Right triangles	Scalene triangles	Isosceles triangles
Similar shapes	Squares	Rectangles
Parallelograms	Rhombuses	Trapezia
Kites	Pentagons	Hexagons
Heptagons	Regular polygons	

Shapes with one, two, three (empty set), four axes of symmetry
Shapes with rotational symmetry
Shapes with equal area, equal perimeter

To make this more practical, we have made 2"×2" cards, each with a different polygon on it. Using a large Venn diagram or Karnaugh diagram, the cards can be sorted into various subsets. A great deal of good exercise in area calculation is involved in sorting into shapes of equal area. Working out perimeters may require the theorem of Pythagoras. When the length of three different line segments is known (Figure 2), all possible perimeters can easily be worked out. Does the figure with largest perimeter have largest area, and vice versa? Is there any relationship between perimeter and area of all shapes?

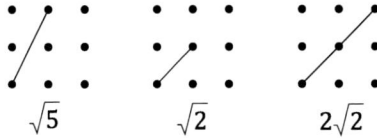

$$\sqrt{5} \qquad \sqrt{2} \qquad 2\sqrt{2}$$

Figure 2

Using the simplest shapes (those with interior angles one right angle, three right angles, or 45°), the interior angle sum of polygons can be calculated.

With a little simple trigonometry (Figure 3), the interior angle sum of the others can be checked:

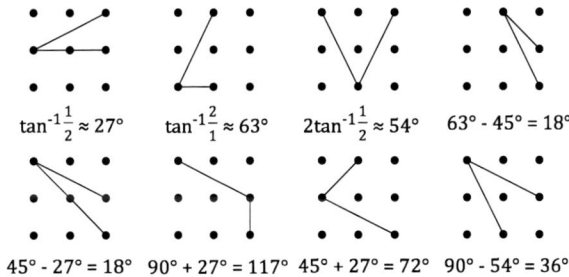

$$\tan^{-1}\frac{1}{2} \approx 27° \qquad \tan^{-1}\frac{2}{1} \approx 63° \qquad 2\tan^{-1}\frac{1}{2} \approx 54° \qquad 63° - 45° = 18°$$

$$45° - 27° = 18° \qquad 90° + 27° = 117° \qquad 45° + 27° = 72° \qquad 90° - 54° = 36°$$

Figure 3

Checking for congruence is best done by rotation of the shape, and superposition. Checking for similarity is trickier. Are the triangles of Figure 4 similar, or are the quadrilaterals of Figure 5?

 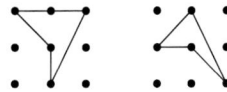

Figure 4   Figure 5

A diagram showing all the 68 polygons we have discovered is at Figure 6. Shapes which can be transformed into those below by reflection of rotation have not been included.

Triangles

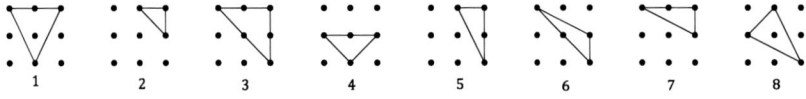

Quadrilaterals

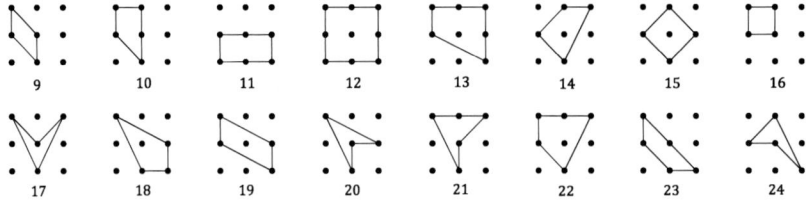

Pentagons

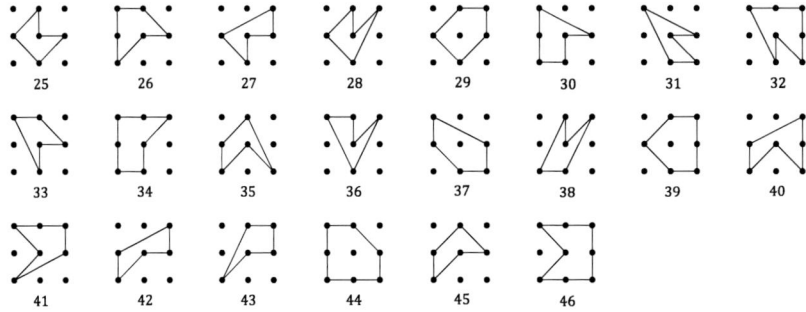

Hexagons

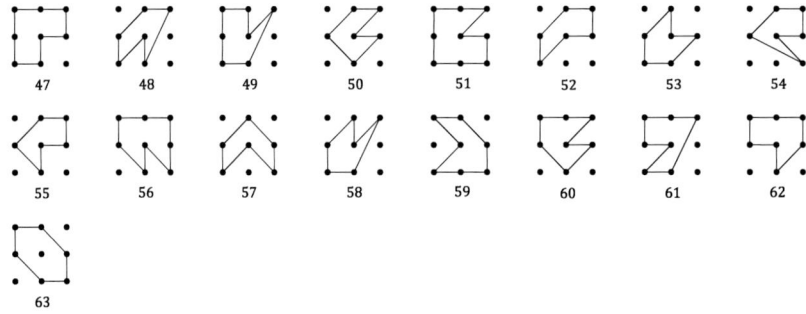

Heptagons

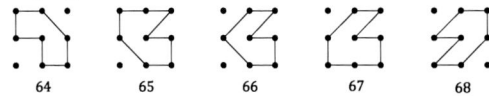

Shape 5 was drawn incorrectly in the original and has been corrected here (*Eds*).

2.2 James A. Dunn on 'Pick's theorem and the hunt for polygons'
Vol. 4, no. 2 (March 1975), pp 6-7

Pick's theorem is now well-known. It is written: $A = \frac{1}{2}P + I - 1$, where A represents the area of a polygon made on a nailboard with a rubber band, P represents the number of nails touched by the rubber band, that is the number of nails on the perimeter; I represents the number of nails inside the shape, untouched by the rubber band. It might be interesting to begin by assuming that this relation is known and investigating its implications. I am not suggesting that it be suddenly pulled out of a hat. Whatever one's general view about doing that, this is much too nice a problem for it; and it has, at an elementary level, the very pleasant and unusual quality of surprise. Its potential as an opportunity for investigation and problem-solving processes is well-known and documented. But after it has been established, with whatever rigour seems appropriate, then the direction of attack might be changed. The relation contains the variables (A,P,I) and the first problem is to decide which one to fiddle with. It's possible to start with any one of the three and produce similar arguments. Starting with P, a bit of reflection suggests that, if there's to be a shape at all at least three nails are needed on the perimeter, also P must be a whole number, so:

(i) $P \in \{3,4,5 \ldots\}$,

(ii) $I \in \{0,1,2 \ldots\}$.

Taking the smallest value of each, i.e. $P = 3$, $I = 0$, and substituting in the relation gives $A = \frac{1}{2}$, and a little bit of thought suggests:

(iii) $A \in \left\{\frac{1}{2},\frac{2}{2},\frac{3}{2},\frac{4}{2},\frac{5}{2} \ldots\right\}$.

With this in mind, returning to the relation $A = \frac{1}{2}P + I - 1$, it is suddenly trivially obvious that when P is even, A is integral. Since this work is being done on a nailboard, the shapes that can be made are more interesting than the formulae and numbers associated with them, although the manipulation of the formula has suggested a problem, "What shapes are possible on a nailboard when A is known?"

(1) Begin with $A = \frac{1}{2}$, then $P + 2I = 3$ and the only possible solution of this is $P = 3, I = 0$. So $(A, P, I) = \left(\frac{1}{2}, 3, 0\right)$. This shape must be a triangle on a unit base and there is an infinite set of them. Some examples are shown in Figure 1.

Figure 1

(2) $A = 1$, then $P + 2I = 4$ and the only possible solution of this is $P = 4, I = 0$. So $(A, P, I) = (1,4,0)$. For this I found two kinds of triangle and a parallelogram. See Figure 2. Have I missed any?

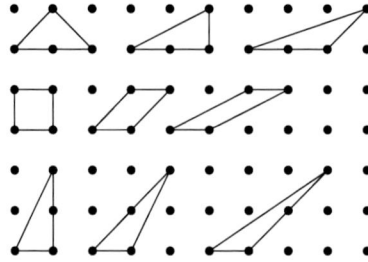

Figure 2

(3) $A = 2$, then $P + 2I = 5$ and this has two solutions $(P, I) = (3,1)$ or $(5,0)$. So $(A, P, I) = (2,3,1)$ or $(2,5,0)$. What shapes are possible? How does it go on? I had just about reached this stage when William Ewbank's article, 'Polygon hunt' was published. The sets of polygons on a nine-pin board were wide-open for a Pick-style analysis, so I wrote the values of P, I, and A for each triangle shown there. I was surprised to find only one of them with area 2, as I could visualize two (see Figure 1). This led me to a mistake. Triangles 5 and 7 are congruent, and so 7 can be replaced with the missing one. There are still some interesting points for discussion. Why can the hexagons alone include cases where $P = 9$? In fact the range of values of P for each shape can be discussed. Would Pick's theorem assist in the creation of those polygons, and would it provide conviction as to when they have all been found?

James Dunn was not alone in noticing the rogue Triangle 5 and identifying the correct version.

5

Six more polygons were also found by readers, taking the total to 74 (*Eds*).

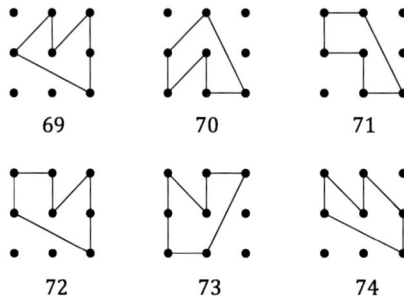

69   70   71

72   73   74

2.3 'A geoboard activity and its possible development',
James K Bidwell and David Hale on geoboard areas,
Vol. 12, no. 3 (May 1983), pp 30-31

The first part of the article gives a brief account of some work done by pupils at Willow Brook Primary School, Keyworth, Nottingham. The children concerned, 10 and 11 year olds, worked in two small groups to produce a collection of polygonal shapes on 9-pin geoboards. The second part of the article takes a critical look at the background to the children's work and suggests some ways in which the activity might be systematically developed and extended.

In the classroom

In one of the two groups, each child initially made a simple polygon on the geoboard and described some of its properties. Their thoughts were then directed towards area by the teacher asking who had made the biggest shape. In the course of the discussion which followed there was some uncertainty about this shape.

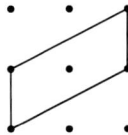

"It's the same as two squares." "A bit less than two." "You can't find it exactly." Agreement in the group was eventually reached by considering the two triangles outside the parallelogram.

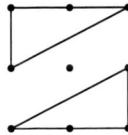

Some could immediately see that the two triangles together accounted for two whole squares. One girl explained that if you cut off a small triangle it could be used to complete a square. Then another pupil pointed out that the original parallelogram could be split into two triangles; so there was no doubt that its area was two squares.

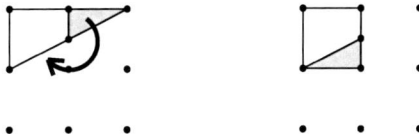

The teacher then asked which other "2 square shapes" could be made and the children worked at this for a while, drawing their shapes on paper prepared with blocks of 9 dots. They then moved naturally to making and recording 1

square shapes, then 3 square shapes and 4 square shapes. Someone discovered that shapes involving ½ squares were possible and eventually every child had a personal collection of shapes, classified by area.

The individual collections were then combined into one. This was achieved by giving all the shapes of a particular area to a pair of pupils and asking them to find out how many were different. The recognition of pairs of congruent shapes caused some difficulty but was largely achieved by the pupils without the intervention of the teacher.

In the other group, pupils also made their own simple polygon initially but this time the teacher asked who had made a shape with the most sides. This introduction led subsequently to each child making a collection of shapes classified by the number of sides. At one stage the teacher asked what was the biggest number of sides possible for a shape on a 9-pin geoboard. An instant answer was 9 (the same as the number of pins), then 6 because someone had made the shape below. There was great excitement when a boy produced a seven-sided shape.

six sides   seven sides

As with the first group, the activity drew to a close with the amalgamation of pupils' individual collections of shapes. In this case, the recognition and elimination of congruent shapes seemed to present more difficulty. In both groups there was apparently little interest in or response to the question – "Have we found all the shapes of a particular type?" However, the first group agreed that their collection of ½, 3½ and 4 square shapes were complete.

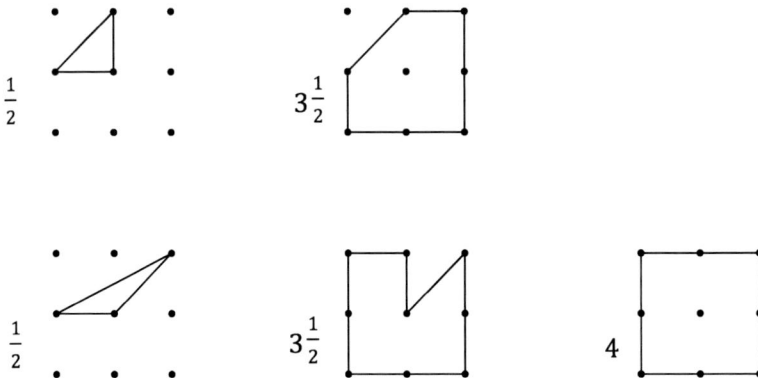

$\frac{1}{2}$

$3\frac{1}{2}$

$\frac{1}{2}$

$3\frac{1}{2}$

4

Drawing the shapes made on the geoboards was not a problem: the prepared paper clearly helped some of the weaker pupils. Curiously, the remaking of a shape from its drawing sometimes posed a difficulty. This occurred when, for example, a pupil wanted to check on the area of a shape he had previously made and drawn.

There were the usual interesting and fruitful discussions about what shapes were acceptable as polygons and what we meant by different shapes. The work concluded with each group looking at and commenting on the connection of shapes produced by the other group.

Analysis and development

The children's (apparently) random production of different polygons led to two immediate related questions. How can they be systematically produced? How many are there? It is clear from Pick's theorem on areas that the smallest possible area is ½ and that this can be formed by using exactly 3 pins. This can clearly be done in exactly two ways, which suggests that other polygons can be systematically produced by adding triangles of area ½ in all possible ways: that is, by joining congruent sides on the geoboard. The 9-pin constraint makes the task relatively simple, provided congruent polygons can be readily detected. (A knowledge of reflections is particularly useful.)

Proceeding to add areas of ½ in all possible ways eventually produced the following numbers of incongruent polygons:

Area	Number found systematically	Number children found
½	2	2
1	8	7
1½	15	6
2	23	15
2½	15	10
3	8	8
3½	2	2
4	1	1
Totals	74	51

Notice the discrepancies between the children's production with no verbalised system and the systematic production. It seemed easier to produce those with large area than those of small area. Why? One of the difficulties with polygons of area 1½, 2 and 2½ is that many of them have very similar shapes which make incongruency hard to see. For example:

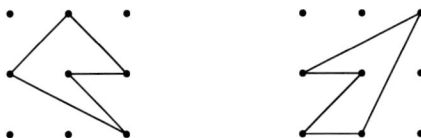

Do you have to add just ½ areas to produce new polygons? Could we combine polygons of areas 1 in all possible ways to produce all polygons of area 2? Or could we combine areas of 1 and 1½ to produce polygons of area 2½? Does this method leave out some possibilities? What do the children think? What do you think?

Another systematic approach is to start with the maximum area of 4 and delete areas of ½ in all possible ways. This may be more satisfactory to many children. If carefully done it will necessarily result in exactly the same collection as the addition method produced. To aid in the work, cut-outs of area ½ triangles are helpful although the whole thing can be easily done using an elastic band and recording each new polygon on a 9-dot grid.

There is of course the other question: "How many are there?" or "Does the systematic method produce them all?" This is an entirely different intellectual problem for children. Although it seems 'clear' to anyone who has generated the polygons in the above methods that this produces all of them – this is no proof. In fact, what proof is acceptable to young children? Is not a systematic approach sufficient for them?

Suppose there was a polygon not produced in this way. Could we not add an area of ½ to it? Would this not lead to area 4 eventually and hence a contradiction? Is this a satisfactory proof? When children are ready to 'analyse' the situation it would seem easy to derive Pick's formula for area from a careful study of polygons with the same area. Concentration on the number of pins on the perimeter (x) and the number of pins inside the polygon (y), should lead to the formula

$$\text{Area} = \frac{1}{2}x + y - 1.$$

Clearly other 'analyses' are possible. Here is a table of sides against areas for polygons on the 9-pin board:

		Area of polygon								
		½	1	1½	2	2½	3	3½	4	Total
	3	2	3	1	2					8
	4		5	3	5	1	1		1	16
No. of sides	**5**			11	5	4	2	1		23
	6				11	7	3	1		22
	7					3	2			5
Total		2	8	15	23	15	8	2	1	74

What kinds of information can be gleaned from this table? If your children are adventurous, they will be interested in different related investigations. How can we generalise? One way is to move to the 16-pin board. This opens the way for a very large number of polygons. A partial investigation reveals the following incomplete table:

Area	Number of polygons
½	4
1	20
.	.
.	.
.	.
8	51
8½	4
9	1

And the number of polygons of area 8 may be even larger. One result is a denial of the symmetry pattern suggested in the first table. A second result is the large increase in the number of polygons of small areas compared with the 9-pin board. This avenue of generalisation seems inappropriate for children on a systematic exploration.

Another way to generalise is to try a triangular geoboard. Although this is only an isometric view of part of the square board, it will not seem so to the child. Thus it offers a new investigation that can be either informally or systematically explored. Because of the isometric deformation, fewer incongruent polygons exist than may be expected. For example, the four triangles of area ½ formed on the 16-pin square board transform into just three triangles on the corresponding 10-pin triangular board. Children may also verify that Pick's theorem works on this geoboard, where each equilateral triangle has area ½.

2.4 Nigel Corps, Paula Ewbank, Ian Simon, Maria Summerfield, Tania Thorpe and Joanna Twist, 14-year olds 'Hunting triangles'
Vol. 12, no. 5 (November 1983), pp 6-9; extract

Getting started

If we were given a number of points symmetrically placed on the circumference of a circle, how many different-shaped triangles could be made by joining together sets of three points? Two triangles were not to be counted as different if one could be changed into the other by a rotation or a reflection (or perhaps a combination of the two).

We were given some duplicated sheets such as this one for the case when there are five points on the circle.

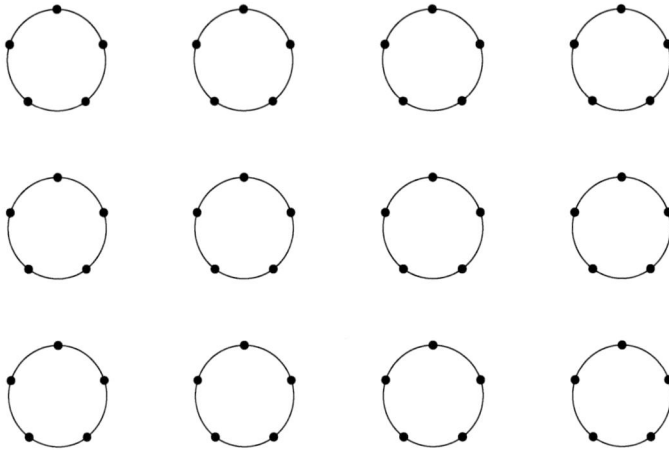

We drew as many different triangles as we could. Our homework for that week was to find as many different ones as we could for the cases in which the number of points, *m*, was 7, 8, 9, 10, 11 and 12. Our results are listed below.

No. of points on the circle, *m*	No. of different triangles, *n*
3	1
4	1
5	2
6	3
7	4
8	5
9	7
10	8
11	10
12	12

This figure shows the triangles we found in the case $m = 12$.

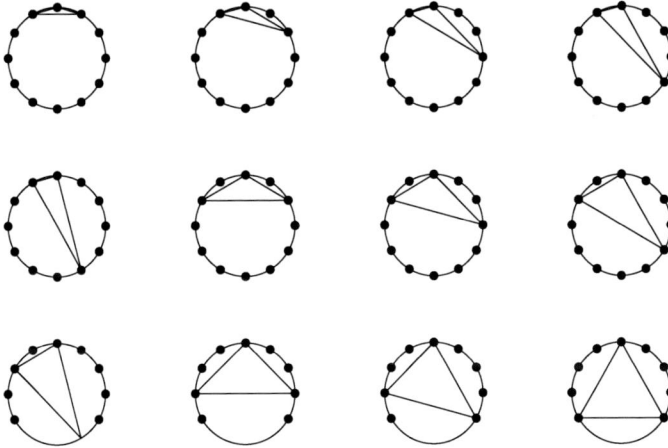

Analysing the results

The following week we met again and compared results. There was some disagreement on a few of our findings, but we compared diagrams and soon ironed out the discrepancies. Our teacher asked us how we would have dealt with the discrepancies if we had been discussing the problem over the telephone and therefore been unable to compare diagrams. Ian suggested counting the points between vertices of each triangle. We discussed the idea and decided to count the spaces between vertices instead. Each triangle could therefore be described by a set of three numbers. We began calling these sets 'triples'. The three numbers in each triple would, of course, add up to m, the number of points on the circle. We were then able to find the number of triangles in the case $m = 13$ by simply listing the triples. We found 14 triples: $(1, 1, 11)$, $(1, 2, 10)$, $(1, 3, 9)$, $(1, 4, 8)$, $(1, 5, 7)$, $(1, 6, 6)$, $(2, 2, 9)$, $(2, 3, 8)$, $(2, 4, 7)$, $(2, 5, 6)$, $(3, 3, 7)$, $(3, 4, 6)$, $(3, 5, 5)$, $(4, 4, 5)$. So there were 14 different triangles. We realised that we could use a computer to find triples (and hence triangles) without any drawing. Two of us wrote a program (i.e. wrote code; *Eds.*) to count the number of possible triples for any chosen value of m. The others found the number of triangles in the cases $m = 14$ to $m = 20$ by listing triples. We found the number of different triangles, n, for values of m up to 32.

m	n	m	n	m	n	m	n
3	1	11	10	19	30	27	61
4	1	12	12	20	33	28	65
5	2	13	14	21	37	29	70
6	3	14	16	22	40	30	75
7	4	15	19	23	44	31	80
8	5	16	21	24	48	32	85
9	7	17	24	25	52		
10	8	18	27	26	56		

Searching for pattern

We then started to look for a pattern. One of the group noticed that if we considered only even values of m then a pattern could be seen in the 'jumps' from one value of n to the next. Our teacher pointed out that values of m which were multiples of 3 were different from the rest because in those cases an equilateral triangle would be possible.

We thought it would be a good idea to concentrate on values of m which were even and multiples of 3, i.e. multiples of 6. The results in these cases are listed below.

No. of points, m	No. of triangles, n
6	3
12	12
18	27
24	48
30	75

After a discussion, we found a rule which worked in these cases. The rule was this: the number of triangles was found by squaring the number of points and then dividing by 12. Hence

$$n = \frac{m^2}{12},$$

for any value of m of the form $6K$ (K a non-negative integer, of course). We then had to try to discover a rule which would apply for other values of m. Tania came up with the idea of examining every sixth result beginning with the case $m = 3$. These were our results.

No. of points, m	No. of triangles, n
3	1
9	7
15	19
21	37
27	61

We soon hit upon the rule $n = \frac{m^2}{12}$, for any value of m of the form $6K + 3$.

It seemed sensible to pursue Tania's idea and look at every sixth value of m, starting with $m = 1$, then with $m = 2$, then with $m = 4$ and finally with $m = 5$. We followed this plan and discovered a rule in each case. These rules, together with the two previously discovered, gave us a six-part formula which we knew worked for any value of m up to 32. It was this:

If m is of the form $6K$, then $n = \dfrac{m^2}{12}$.

If m is of the form $6K + 1$, then $n = \dfrac{m^2 - 1}{12}$.

If m is of the form $6K + 2$, then $n = \dfrac{m^2 - 4}{12}$.

If m is of the form $6K + 3$, then $n = \dfrac{m^2 + 3}{12}$.

If m is of the form $6K + 4$, then $n = \dfrac{m^2 - 4}{12}$.

If m is of the form $6K + 5$, then $n = \dfrac{m^2 - 1}{12}$.

Proof

Our teacher then asked us if we thought that we had proved that this six-part rule would give us the number of different triangles for any value of m. We were all confident that we had but he said that this was not so and that all we had done was to find a rule (made up of several parts) which fitted the known results for $m = 3, 4, 5, ... \ 32$. Who was to say that this rule would work for values of m greater than 32?

One of the group suggested using our rule to predict the number of triangles in the case, say, $m = 50$ or $m = 77$ and then using the computer to check our predictions. We did this several times and were proved correct every time.

This confirmed our belief that our rule was correct but we still felt that we needed a proof. Our teacher proved to us that our rule was correct when m was of the form $6K + 1$. We found the beginning of the proof difficult to understand but gained confidence as we went along. At the end we felt that we had understood it. Here it is. Let us list all possible triples in the cases $m = 7$ and $m = 13$.

$m = 7$		$m = 13$			
$(1,1,5)$	$(2,2,3)$	$(1,1,11)$	$(2,2,9)$	$(3,3,7)$	$(4,4,5)$
$(1,2,4)$		$(1,2,10)$	$(2,3,8)$	$(3,4,6)$	
$(1,3,3)$		$(1,3,9)$	$(2,4,7)$	$(3,5,5)$	
		$(1,4,8)$	$(2,5,6)$		
		$(1,5,7)$			
		$(1,6,6)$			

Now in the general case $m = 6K + 1$, the triples in the first column will be

$(1,1,6K - 1)$
$(1,2,6K - 2)$
.
.
.
$(1,3K,3K)$ giving $3K$ triples in the first column.

In the second column we will have

$(2, 2, 6K - 3)$
$(2, 3, 6K - 4)$
.
.
.
$(2, 3K - 1, 3K)$ i.e $3K - 2$ triples in the second column.

In the third column we will have

$(3, 3, 6K - 5)$
$(3, 4, 6K - 6)$
.
.
.
$(3, 3K - 1, 3K - 1)$ i.e. $3K - 3$ triples.

We can continue in this way until we reach the last two columns in which the triples will be

$(2K - 1, 2K - 1, 2K - 3)$
$(2K - 1, 2K, 2K + 2)$
$(2K - 1, 2K + 1, 2K + 1)$ i.e 3 triples.

Hence the number of triples (and therefore the number of distinct triangles) is

$$n = 3K + (3K - 2) + (3K - 3) + (3K - 5) + \cdots + 3 + 1$$
$$= [3K + (3K - 3) + (3K - 6) + \cdots + 3] + [(3K - 2) + (3K - 5) + \cdots + 1]$$
$$= \frac{K}{2}(3K + 3) + \frac{K}{2}(3K - 1)$$
$$= K(3K + 1).$$

But $m = 6K + 1 \Rightarrow K = \dfrac{m - 1}{6}$.

So, $n = \left(\dfrac{m - 1}{6}\right)\left[\left(\dfrac{m - 1}{2}\right) + 1\right] = \dfrac{m^2 - 1}{12}$.

We were sure that the other parts of the rule could be proved in a similar way. At last we felt satisfied that the problem was solved.

2.5 'Seven pins' by Dave Kirkby
Vol. 15, no. 2 (March 1986), pp 14-15

[The cards below show] the different shapes that can be made on a seven-pin geoboard. [...] They have provided me with a source for developing many geometrical ideas appropriate for children in both primary and secondary schools.

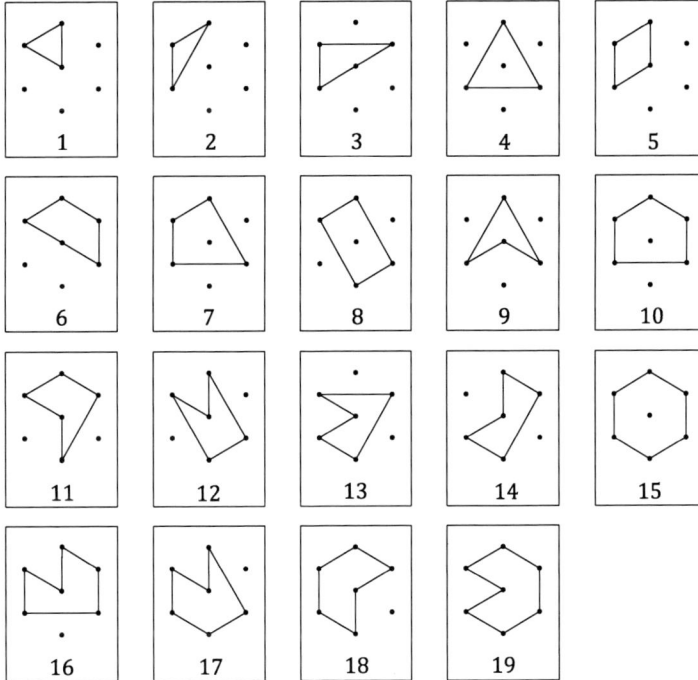

Angles

1. Sorting activities

Equal angles. Which of the shapes have:
    (a) all angles different?
    (b) two, three, four, five, six equal angles?

Right angles. Which shapes contain:
    (a) one right angle? (b) two, three, four right angles?

Other types of angle. Which shapes contain:
    (a) all acute angles?
    (b) an obtuse angle?
    (c) two obtuse angles?
    (d) all obtuse angles?
    (e) a reflex angle?

2. Angle between any two lines

What is the angle between the following pairs of lines?

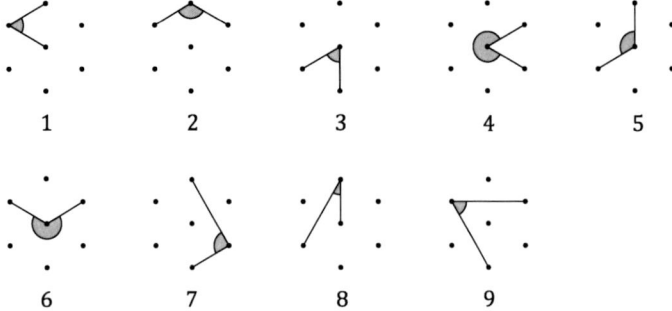

     1   2   3   4   5

     6   7   8   9

3. Angles of the polygons

From 2, the angles of each polygon can now be calculated, e.g. consider the five quadrilaterals

Shape	Angles
5	60°, 60°, 120°, 120°
6	60°, 60°, 120°, 120°
7	120°, 90°, 90°, 60°
8	90°, 90°, 90°, 90°
9	30°, 30°, 60°, 240°

What are the sums of the angles of these quadrilaterals?
What do you notice about the opposite angles of 6, 7 and 8?
N.B. If the angles of each of the other polygons are calculated, this provides a starting point for work on the angle sum of polygons.

Sides

1. Sorting Activities

Equal sides. Which of the shapes have:
     (a) all sides of different length?
     (b) two, three, four, five, six equal sides?

Parallel sides. Which of the shapes have:
     (a) one pair
     (b) two pairs
     (c) three pairs of parallel sides?

Perpendicular sides. Which of the shapes have a pair of perpendicular sides.

2. Lengths of sides

Three different lengths of sides exist.

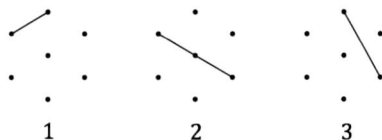

Let the distance between two adjacent dots be 1 unit. Then,

    length of line (1) is 1 unit,
    length of line (2) is 2 units,

and by Pythagoras,

    length of line (3) is $\sqrt{3}$ units.

The lengths of sides of each polygon can now be stated [and the perimeters found], e.g. consider the four triangles.

Shape	Sides	Perimeter
1	1, 1, 1	3
2	1, 1, $\sqrt{3}$	$2 + \sqrt{3}$
3	1, 2, $\sqrt{3}$	$3 + \sqrt{3}$
4	$\sqrt{3}, \sqrt{3}, \sqrt{3}$	$3\sqrt{3}$

If the lengths of sides of each shape are tabulated, then: Which shapes have all sides the same length? Which shapes have all sides of different length? Which shape has (a) the greatest perimeter (b) the smallest perimeter? Which shapes have equal perimeters?

Properties of shapes

1. Written description

The teacher chooses a card at random. Pupils are required to write five sentences to describe the properties of the shape on the card.

2. Spoken description

An activity for two children. The nineteen cards are placed face-up on the table. One child 'thinks of a shape' without touching it. He then describes the shape on this card to the other child e.g. "I am thinking of a shape which is a pentagon with one pair of parallel sides. It also contains two right angles and a reflex angle." The other child has to guess which card has been chosen.

3. Guessing game

A game for two players. Place the nineteen cards face up on the table. One player thinks of a card without touching it. The other player has to guess the card by asking a series of questions to which the reply can only be "Yes" or "No". Examples of questions might be: "Does it have a right angle?" "Does it have more than four sides?" "Does it have an axis of symmetry?" When the chosen shape has been discovered, the numbers of required guesses are recorded. Then the roles are reversed. The player who requires the fewer questions wins.

Area

Consider the area of each shape. Use the equilateral triangle as the unit of area i.e. the area of shape 1 is one unit. From this base the areas of shapes 5, 6, 15, 18 and 19 can be established.

If, then, the area of shape 2 is established as one unit (one half of two units) the areas of all the other shapes can be found. Which shape has maximum area? Which shape has minimum area? Which shapes have equal areas?

Symmetry

Sort the shapes into two sets according to whether or not they have an axis of symmetry. Consider the shapes which do have an axis of symmetry.

Which of these have 3 axes of symmetry? What do you notice about them?

Which have 2 axes of symmetry? What do you notice about them?

Which shapes have rotational symmetry?

2.6 'Forming and adjusting conjectures: Perimeter and pattern blocks', by Bonnie H. Litwiller and David R. Duncan
Vol. 26, no. 2 (March 1997), pp 22-25

Teachers are always looking for novel ways to incorporate geometric and measurement concepts into their classrooms. One such way is to use the following four blocks from a set of pattern blocks. Figure 1 displays these blocks, drawn on an isometric geoboard for ease in reporting and determining the measures of the sides.

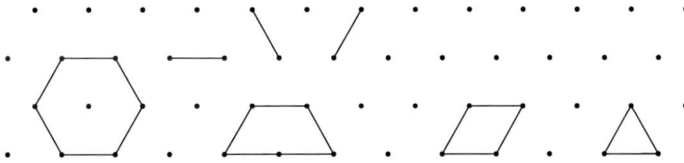

Fig. 1

As an initial activity, use two of each pattern block to form various polygonal arrangements, and find their perimeters by counting. Figure 2 displays several such examples. Each of these figures can be considered a 'train'; that is, one can traverse the arrangement from one end to the other without repeating any block. In each train, the blocks are numbered in the order they can be traversed. The order may be reversed.

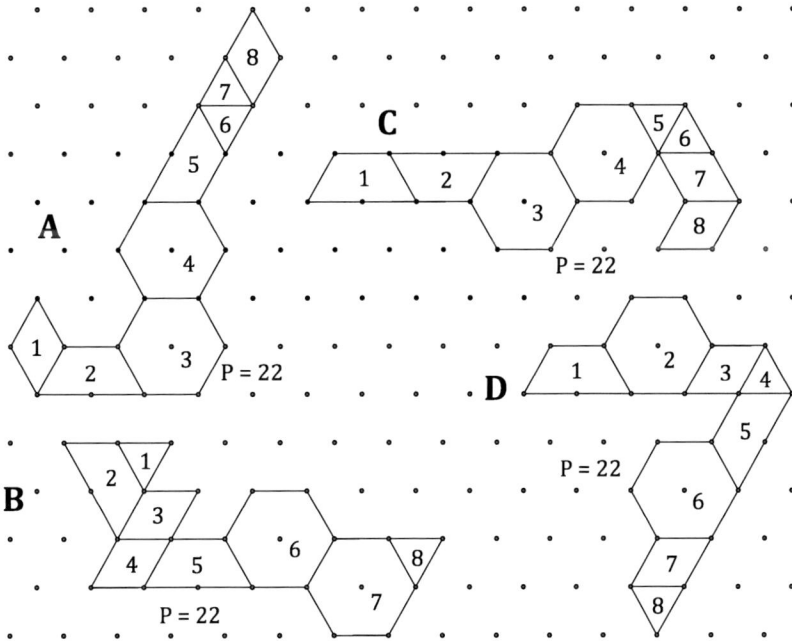

Fig. 2

Observe that the perimeter of each arrangement is 22. How might this constant perimeter be explained?

From Figure 1, the perimeters of the blocks are:

hexagon = 6
trapezoid = 5
rhombus = 4
triangle = 3.

For a given train, each of the blocks 2 through 7 contribute 2 less than its original perimeter to the train's perimeter. One side is 'lost' when we enter the block; another side is lost when we exit the block. Blocks 1 and 8 contribute 1 less than their original perimeters because they are either entered or exited, but not both, in the traversing process. Based upon these four examples, your students may conjecture that any train containing two hexagons, two trapezoids, two rhombi, and two triangles (in which each block-pair intersection has only one unit of length) should have perimeter:

$$2[(6-2)+(5-2)+(4-2)+(3-2)]+2 = 2[4+3+2+1]+2 = 22.$$

Support this conjecture by making additional trains, using two of each block. A general conjecture can be made.

Suppose that a train contains: h hexagons T trapezoids r rhombi t triangles and that each block-pair intersection has length one; then, the perimeter of this train is $[h(6-2)+T(5-2)+r(4-2)+t(3-2)]+2 = 4h-3T+2r+t+2$.

For example the train of Fig. 3 has $h = 1$, $T = 3$, $r = 3$ and $t = 2$. Have your students count that its perimeter is 23; this agrees with the predicted value of

$$[1 \times 4 + 3 \times 3 + 3 \times 2 + 2 \times 1] + 2 = 23.$$

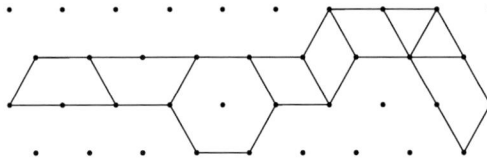

Fig. 3

What happens if the train of Fig. 3 is modified as in Fig. 4 Observe that two of the trapezoids now form a hexagon, so the perimeter should now be:

$$[2 \times 4 + 1 \times 3 + 3 \times 2 + 2 \times 1] + 2 = 21.$$

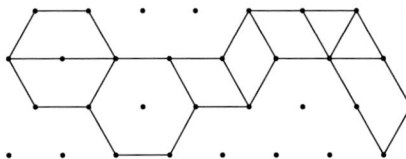

Fig. 4

Verify this is true by counting the number of units to find the perimeter. What happens if a non-train arrangement is formed? Re-arranging the blocks of Fig. 3 yields the arrangement displayed in Fig. 5.

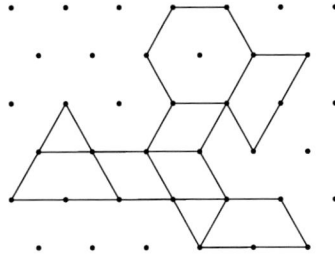

Fig. 5

Again, the one-unit intersection restriction is used. By direct counting, the perimeter is 23, the same as the arrangement in Fig. 3. The general conjecture $P = 4h + 3T + 2r + t + 2$ still holds even though the rationale used in its development (for trains) no longer applies. How might this be explained? Could the notion of two trains be used?

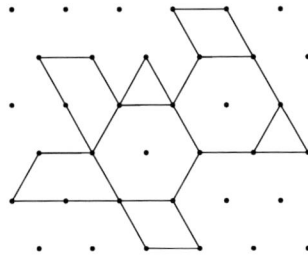

Fig. 6

Another non-train is shown in Fig. 6. Its perimeter can be counted to be 22 which agrees with the general conjecture of

$$[2 \times 4 + 2 \times 3 + 2 \times 2 + 2 \times 1] + 2 = 22.$$

Challenges for teachers and students

1. Consider the following two arrangements:

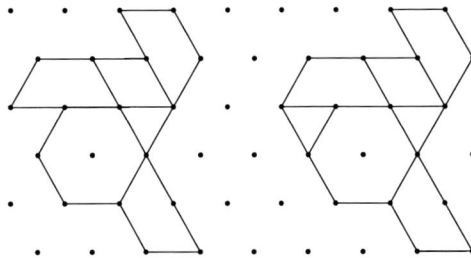

The general conjecture fails to correctly predict the perimeters of the two arrangements. Note, however, that each arrangement has at least one tessellated vertex of blocks in its interior. How can the presence of these tessellated points be used to adjust the general conjecture?

2. The following two arrangements are 'punctured'; that is, there are holes in their interiors. How can previous conjectures be adjusted to account for these punctures?

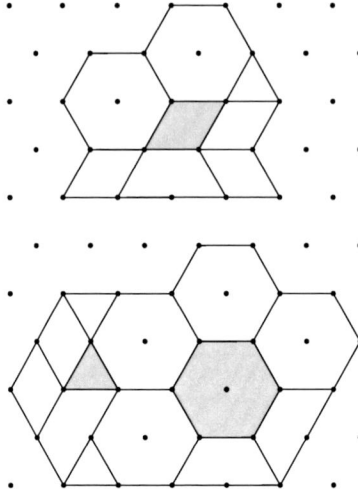

2.7 'An investigation that takes some unusual turns', by Bob Shafee
Vol. 20, no. 2 (March 1991), pp 34-37; extract

WHAT MAKES A RICH MATHEMATICAL ACTIVITY?
- It must be accessible to everyone at the start.
- It needs to allow further challenges and be extendible.
- It should invite children to make decisions.
- It should involve children in speculating, hypothesis making and testing, proving or explaining, reflecting, interpreting.
- It should not restrict pupils from searching in other directions.
- It should promote discussion and communication.
- It should encourage originality/ invention.
- It should encourage "what if" and "what if not" questions.
- It should have an element of surprise.
- It should be enjoyable

Such qualities from an investigation are always to be sought, but not easily found. How sweet it is when they occur unexpectedly for it is then that you can really share that element of discovery.

The problem

This investigation is one I used with my first year group and can be stated as follows:

From a fixed starting point on a square grid, what points can you reach by moving 3 squares, or "steps", in any of the four directions?

Investigate further.

Starting point

It did not take them long to realise that the points form a particular pattern and provide, as a result, a familiar number sequence.

Number of steps, n	Number of points reached, P_n
1	4
2	9
3	16
n	$(n+1)^2$

$$P_n = (n+1)^2$$

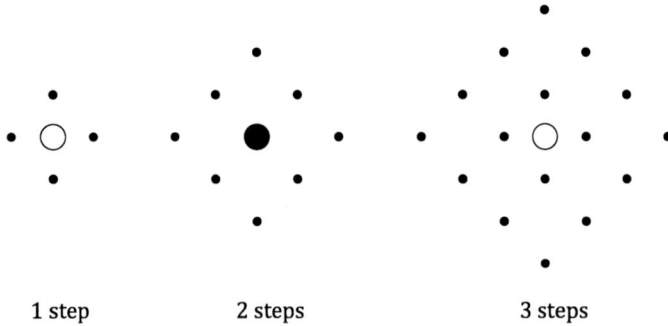

| 1 step | 2 steps | 3 steps |

Prediction is a readily accessible process, and easy to check too. The algebraic generalisation seemed inappropriate for this age group, but Claire put it quite clearly in words:

The number of moves made by each amount of steps is the square of the number one higher than that number. e.g.
1 step = 4 moves, 2 squared = 4
2 steps = 9 moves, 3 squared = 9

If the dots are viewed in rows the following analysis is apparent.

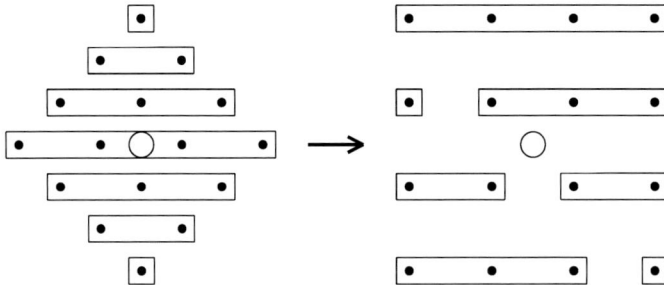

i.e. $(1 + 2 + 3) + (1 + 2 + 3 + 4) = P_3$
or $T_3 + T_4 = P_3$, where T_3 is the third triangular number.

In general then, $T_n + T_{n+1} = P_n$, which can be used to confirm the rule:

$$T_n + T_{n+1} = \frac{1}{2}n(n + 1) + \frac{1}{2}(n + 1)(n + 2)$$

$$= \frac{1}{2}(n + 1)(2n + 2)$$

$$= (n + 1)^2.$$

Dissecting the dots in a different way leads to a second approach.

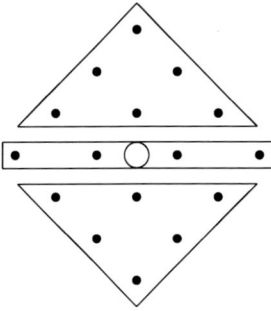

$$2(1 + 2 + 3) + 4 = P_3 \text{ or } 2T_3 + 4 = P_3.$$

This now suggests the general rule:

$$2T_n + (n + 1) = P_n$$

and is confirmed thus:

$$2T_n + (n + 1) = n(n + 1) + (n + 1)$$
$$= (n + 1)(n + 1)$$
$$= (n + 1)^2.$$

How many routes?

Mary looked at how many different ways there are to reach each point, and this was the first unexpected departure from the main theme. The results for the first 3 stages are:

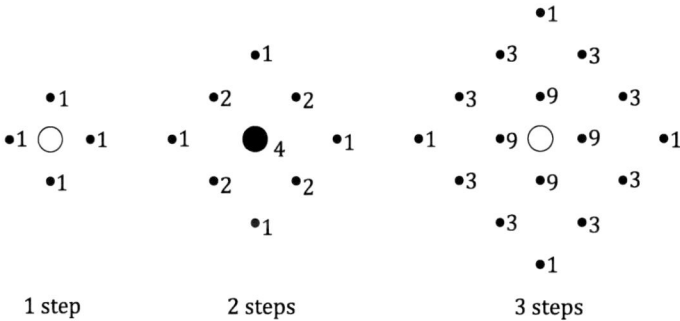

1 step	2 steps	3 steps

and the number patterns look interesting. As a consequence I decided to look a little further into the problem. Tracing out the routes quickly becomes too tedious an exercise, and liable to error, so a more analytical approach is needed. The appearance of the rows of Pascal's triangle round the edge of the grid is a reminder of how this number pattern is sometimes introduced.

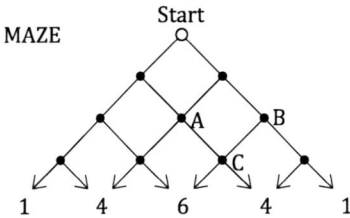

Start at the top of the maze and move downwards each time, either left or right. The number of different routes to each exit point give the rows of the triangle.

Further extension of the triangle using the familiar rule is simply explained by the fact that, in order to reach the point C, for example, it is necessary to pass through either points A or B. Hence the number of routes to C is the sum of the routes to A and B. Now, a Side Step movement is in one of four directions, not

67

two, but the approach is similar. To solve the 4 step problem, superimpose the
grid (X's) onto the 3-step grid (•'s) as shown.

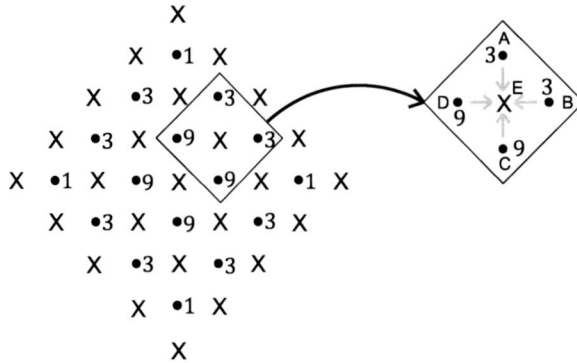

To reach point E in 4 steps requires passing through A, B, C or D in 3 steps. So
this gives the number of steps to E as 3 + 3 + 9 + 9 = 24. In this way the solutions
to the 4 and 5 step problems can be found.

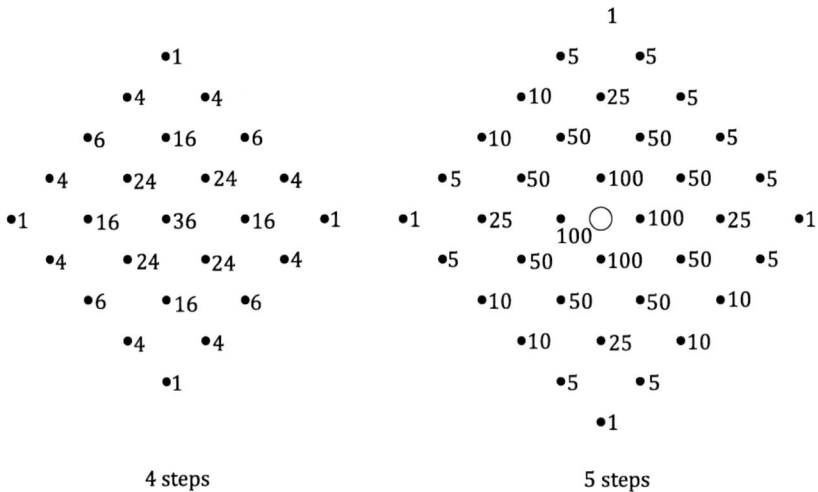

4 steps

5 steps

If these solutions are now regarded as arrays of numbers, stacked one on top of
another, they then form a 3-dimensional matrix.

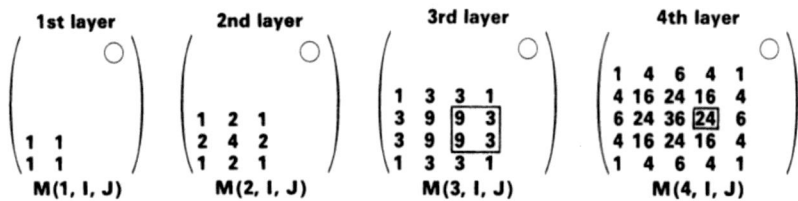

1st layer	2nd layer	3rd layer	4th layer

M(1, I, J)   M(2, I, J)   M(3, I, J)   M(4, I, J)

As observed before 24 = 3 + 3 + 9 + 9

or   M(4,4,3) = M(3,4,3) + M(3,4,2) + M(3,3,2) + M(3,3,3), or in general

$$M(S + 1, I, J) = M(S, I, J) + M(S, I, J - 1) + M(S, I - 1, J - 1) + M(S, I - 1, J).$$

This is equivalent to the defining relation in Pascal's triangle:

$$\binom{n+1}{i} = \binom{n}{i-1} + \binom{n}{i}.$$

A border of zeros to the left and below is sufficient to make the definition universal, in that $I = 1$ and $J = 1$ become valid parameters, and a computer program will readily evaluate subsequent layers as required.

Finally, looking quickly at the total number of routes in each grid gives the following:

Number of steps, n	Number of routes, R_n
1	4
2	16
3	64
4	256

and suggests the general rule, $R_n = 4^n$. This is confirmed by noting that these arrays are multiplication tables using rows from Pascal's triangle.

×	1	3	3	1
1	1	3	3	1
3	3	9	9	3
3	3	9	9	3
1	1	3	3	1

×	2^3
2^3	4^3

Since the nth row of the triangle has a sum 2^n the resulting product is 4^n.

Different grids

Back in the classroom it was suggested trying different shaped grid paper. Triangular paper produces these results, for 1, 2 and 3 steps respectively:

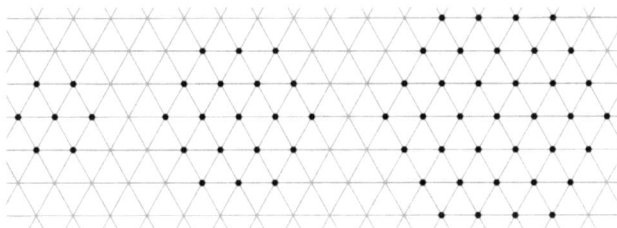

Number of steps, n	1	2	3	4
Number of points reached, P_n	6	19	37	61

Ignoring the first result, the pattern was again clearly outlined by Claire:

I found that each amount of moves went up in multiples of 6.
The multiplier was the same numbers the number of steps.
e.g. 3 steps = 19 + 3 × 6, 4 steps = 37 + 4 × 6

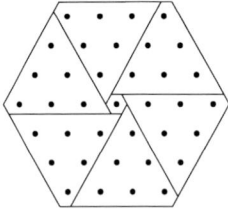

A familiar dissection of the three-step grid leads to $P_3 = 6T_3 + 1$ and in general $P_n = 6T_n + 1$ which gives $P_n = 3n(n + 1) + 1$.

Hexagonal grid paper however provides the most interesting results:

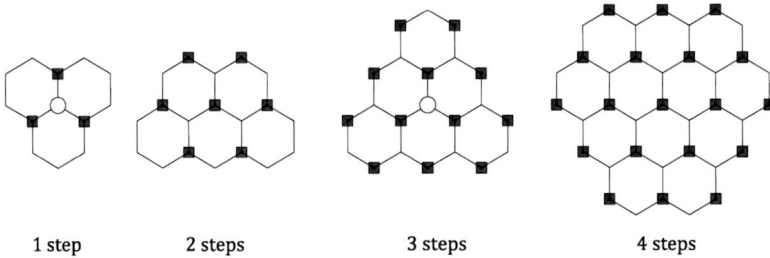

1 step	2 steps	3 steps	4 steps

n	1	2	3	4
P_n	3	7	12	19

Again I was prompted to look into the situation. Dissecting the pattern for 4 steps, as shown in the diagram below, leads to the following breakdown.

$\rightarrow$

n	P_n
1	$3 = 1 + 2$
2	$7 = 2 + 3 + 2$
3	$12 = 2 + 3 + 4 + 3$
4	$19 = 3 + 4 + 5 + 4 + 3$

This can be extended to predict the next few results.

n	P_n
5	$27 = 3 + 4 + 5 + 6 + 5 + 4$
6	$37 = 4 + 5 + 6 + 7 + 6 + 5 + 4$

If these results are now divided into two groups, odd steps and even steps, and the series rearranged slightly, the following can be obtained.

n	P_n
1	$3 = 1 + 2$
3	$12 = 2 + 3 + 3 + 4$
5	$27 = 3 + 4 + 4 + 5 + 5 + 6$
7	$48 = 4 + 5 + 5 + 6 + 6 + 7 + 7 + 8$

n	P_n
2	$7 = 2 + 2 + 3$
4	$19 = 3 + 3 + 4 + 4 + 5$
6	$37 = 4 + 4 + 5 + 5 + 6 + 6 + 7$
8	$61 = 5 + 5 + 6 + 6 + 7 + 7 + 8 + 8 + 9$

Now, considering the odd number results, each term in the series fits together to form three-quarters of a square. Here are the first three:

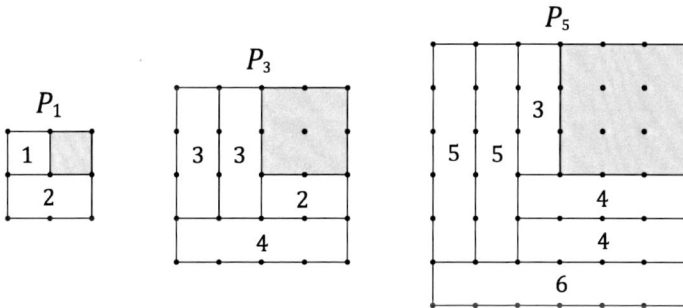

This then leads to the general result for n odd,

$$P_n = \frac{3}{4}(n + 1)^2.$$

The even number results are not quite so helpful when the same process is applied to them.

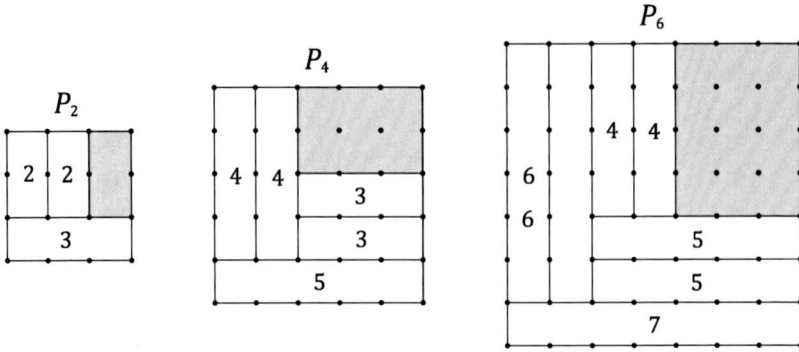

$$P_2 \qquad P_4 \qquad P_6$$

Now this would suggest

$$P_n = (n+1)^2 - \frac{n}{2}\left(\frac{n}{2}+1\right)$$

but one part may be moved (A), and a quarter-square removed (B), like this:

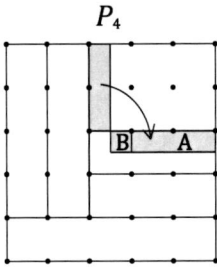

This suggests, that for *n* even,

$$P_n = \frac{3}{4}(n+1)^2 + \frac{1}{4}.$$

2.8 Alan Burns on 'The polygon trail'
Vol. 11, no. 4 (September 1982), pp 8-13; extract

Consider the problem of determining the maximum number of non-congruent figures which can be made on the nine-pin geoboard. The following is an account of an investigation into this problem and endeavours to give a proof for the maximum number of n-sided figures and hence of all figures possible on the board. Firstly, as the simplest case, consider how many different points are possible.

Points

There are nine ways of choosing a single pin from nine:

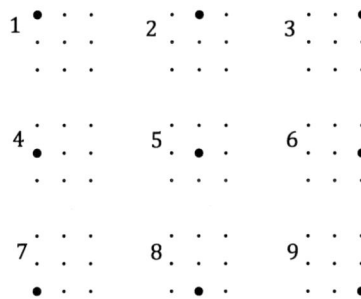

Some of these ways may be regarded as being the same. For instance, Figure 1 is considered to be the same as Figure 3 because the arrangement of pins in 3 can be made into that of 1 simply by turning the board.

Arrangements 1 and 3 are said to be equivalent. Other arrangements which are equivalent to 1 and 3 are 7 and 9. The set of arrangements which are equivalent to 1 will be called the equivalence class of 1. There are four members of this class; 1, 3, 7, 9. The arrangement 2 does not belong to the above class because there is no way of mapping it onto 1. Arrangement 2 has its own equivalence class, and this also has four members; 2, 4, 6, 8. Arrangement 5 does not belong to either of the classes already mentioned, 5 has an equivalence class consisting only of itself. These results can be put in the form of a table.

Equivalence class (EC)	Pin pattern	Number in EC	Members of EC
A1		4	1, 3, 7, 9
B1		4	2, 4, 6, 8
C1		1	5

The equivalence class containing 1 has been labelled A1, that containing 2, B1 and that containing 5, C1. The total number of arrangements in all three classes is nine and this takes up all the possible ways of choosing one pin from nine. This means that there are no arrangements that are not in one of the classes A1, B1 and C1. It has thus been proved that there are only three different 1-pin arrangements possible on the 9-pin board.

Line segments

Just as in the case of points, the two-pin patterns which produce linear figures can be grouped into equivalence classes. The table is given below.

Equivalence class (EC)	Pin pattern	Number in EC	Figure
A2	• • · · · · · · ·	12	— · · · · · · ·
B2	• · • · · · · · ·	6	—— · · · · · ·
C2	• · · · • · · · ·	8	\ · · · · · · · ·
D2	• · · · · · · · •	2	· · · \ · · ·
E2	• · · · · · · • ·	8	· \ · · · · · ·

The total number of arrangements in all of these classes is 36. In order to prove that the above are the only equivalence classes, it remains only to show in how many ways two pins may be chosen from nine. The results of combination theory give the required value quite simply; the number of ways of choosing two pins from nine is:

$$\binom{9}{2} = \frac{9!}{2!\,(9-2)!} = 36.$$

The members of the five equivalence classes already found account for all of these 36 possibilities, and so there can be no more equivalence classes. Thus it has been proved that there are just five different line segments possible on the nine-pin geoboard.

Triangles

In the case of choosing three pins from nine the number of ways of doing so is:

$$\binom{9}{3} = \frac{9!}{3!\,(9-3)!} = 84.$$

Thus the total number of members in all of the equivalence classes must be 84.

Equivalence class (EC)	Pin pattern	Number in EC	Figure
A3	● · · · · · ● · ●	4	(triangle)
B3	· ● · · · · ● · ●	4	(triangle)
C3	· · · ● · · ● · ●	16	(triangle)
D3	· · · · ● · ● · ●	8	(triangle)
E3	· · · ● · · ● ● ·	16	(triangle)
F3	· · · · · ● ● ● ·	16	(triangle)
G3	· · ● · · · ● ● ·	8	(triangle)
H3	· ● · · · ● ● · ·	4	(triangle)

The total number of members in the equivalence classes already found is 76. This is less than the 84 arrangements which are possible. The difference arises because some three-pin patterns do not produce triangles, but straight lines. The equivalence classes which contain these patterns are said to be degenerate; the patterns produce figures which it is possible to make with fewer pins.

	Pin pattern	Number in EC	Figure
a	● · · ● · · ● · ·	6	(line)
b	● · · · ● · · · ●	2	(line)

The total number of members in the equivalence classes which are degenerate is 8. This, together with the 76 previously found makes up the total of 84, and so there can be no more classes. It has been proved that there are just eight different triangles possible on the nine-pin geoboard.

It is noticeable that the degenerate patterns *a* and *b* give rise to the same figures as do B2 and D2 respectively. By a consideration of the case of line segments it is seen that B2 and D2 are the only patterns which give rise to a figure which passes through three points. Also, the number in class B2 is the same as that in *a*; the number in D2 is the same as that in *b*. This gives a method of finding the number of degenerate patterns for any number of pins.

Quadrilaterals

The number of ways of choosing four pins from nine is:

$$\binom{9}{4} = \frac{9!}{4!\,(9-4)!} = 126.$$

The degenerate classes can now be found by a consideration of the results for $n < 4$ pins. It is found that patterns A3, B3, C3, D3 and G3 all give rise to figures which pass through at least four pins; a fourth pin may be added to A3 in any of three positions to give a pattern which will produce the same figure as A3.

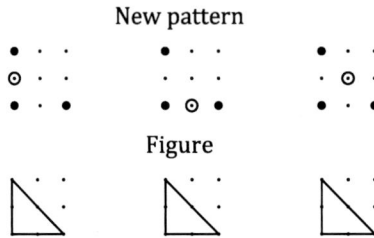

New pattern

Figure

For B3, C3, D3 and G3 there is only one such position for a fourth pin. Thus there will be three degenerate patterns of four pins for each member of A3 and one for each of those in B3, C3, D3 and G3.

		Equivalence class		
A3	B3	C3	D3	G3
		Number in EC		
4	4	16	8	8
		Ways of placing fourth pin		
3	1	1	1	1
		Degenerate patterns		
		Degenerates		
12	4	16	8	8

Pattern A3 gives rise to two different degenerate patterns, one in an equivalence class of eight members, the other in an equivalence class of four members. Hence there is a total of 48 degenerate patterns of four pins.

	Pin pattern	Number in EC	Figures possible
A4		1	
B4		8	
C4		8	
D4		4	
E4		4	
F4		16	
G4		8	
H4		4	
I4		8	
J4		4	
K4		4	
L4		4	
M4		4	
N4		1	

Adding all the members of the 14 equivalence classes found gives a total of 78, this together with the 48 degenerates gives a total of 126, as required. It is seen that in classes D4 and M4, the same pattern gives rise to more than one figure.

It is thus necessary to make sure that all of the possible figures of each pattern has been found. Firstly, it is useful to know a few lemmas which will help in deciding how many figures are possible from a given pattern.

Lemma 1 Re-entrant figures are only possible on the nine-pin geoboard iff the patterns producing them involve the central pin.

Proof

If the central pin is not used, only the perimeter pins are available. The vertices of the figure must then occur at the perimeter of the board. Thus the re-entrant vertex occurs on the perimeter, i.e. it intersects one of the sides of the figure.

These figures are not truly re-entrant. Figure D4 shows that re-entrant figures are possible where the central pin is used.

Lemma 2 If an equivalence class contains a non-re-entrant pattern then only one figure may be produced from that pattern and it will be a non-re-entrant $(n - r)$ figure.

Proof

For a non-re-entrant figure on the nine pin geoboard, the central pin is the only one which may lie within the figure. The diagonals of a $n - r$ figure all lie within the figure.

Suppose that a second figure does exist for a given pin pattern. Not all of the line segments used in the original $n - r$ figure may be used in the second figure. The only other line segments available for the perimeter of the second figure are the diagonals of the original.

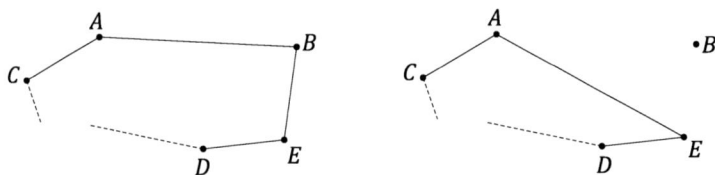

At least one of the diagonals of the original must be used; consider the simplest case of a diagonal joining two next-but-one neighbours as the only diagonal used. Say the diagonal AE above. In order to include this diagonal, the point B must be excluded from the figure. The lines AB and BE are replaced by the single line AE. The number of sides of the new figure is less than the original: the new figure is degenerate and is not in the equivalence class of the original figure.

Lemma 3 The greatest number of figures possible from a re-entrant n-pin arrangement on the nine-pin board is $n - 1$.

Proof

For re-entrance, two adjacent pattern pins on the perimeter of the board must be connected to the central pin. A different re-entrant vertex is produced by each pair of adjacent pins on the perimeter; the number of such pairs is just the number of pins on the perimeter, i.e. $n - 1$. Now, triangles cannot be re-entrant and in the case of quadrilaterals, only D4 and M4 can possibly give rise to more than one figure since they are the only re-entrants. Further, these can give at most three different figures. In the rest of this work it will be assumed that this type of investigation has taken place for all re-entrant pin patterns, and only the different figures from these patterns will be detailed. As an example of a degenerate pattern, consider D4:

	Pin pattern	Figures possible
D4		

There are just two different (non-congruent) figures.

Pentagons

The number of ways of choosing five pins from nine is:

$$\binom{9}{5} = \frac{9!}{5!\,(9-5)!} = 126.$$

The degenerate figures are found from an investigation of quadrilaterals and triangles.

	Equivalence class		
A3	A4	B4	C4

	Number in EC		
4	1	8	8

	Ways of placing remaining pins		
3	4	2	1

	Degenerate patterns		

	Degenerates		
12	4	16	8

Equivalence class	D4	E4	F4	L4
Number in EC	4	4	16	4
Ways of placing remaining pins	1	2	1	1
Degenerate patterns				
Degenerates	4	8	16	4

Both of the patterns A3 and B4 give rise to two distinct degenerate patterns; for A3 one of the degenerate patterns has an equivalence class of eight members and the other four members, for B4 both patterns are in equivalence classes of eight members. There is a total of 72 degenerate figures.

	Pin pattern	Number in EC	Figures possible
A5		4	
B5		8	
C5		4	
D5		1	
E5		8	
F5		8	
G5		4	
H5		4	

I5		4	
J5		4	
K5		8	
L5		1	

The total number of members in the 12 non-degenerate equivalence classes is 58. Together with the 72 degenerate figures this gives a total of 130 possible figures, more than the 126 predicted algebraically.

A miscalculation must have occurred; it could not have been in the number of non-degenerates, the proof of this being their existence in the above diagrams. Something is wrong with the method of finding the degenerates. An over-calculation of four figures has been made. Assuming that no simple arithmetical mistake was made, one of the classes must be non-degenerate. It is seen that D4 is non-degenerate, producing 15.

Having eliminated the four D4 figures, the total of degenerate and non-degenerate figures found is 126 as required. This gives a proof that there are just 23 different pentagonal figures.

It may be useful at this point to give an explanation of exactly how this proof works. Firstly, as many n-pin figures as possible are found using trial and error and the number of non-congruent patterns thus found is taken as a trial value for the number of non-degenerate n-pin patterns possible.

Secondly, the number of degenerate patterns is calculated from an observation of the $r < n$ pin patterns using lemmas 1 to 3.

Finally the total number of degenerates plus the patterns found is compared with the algebraic calculation of patterns possible and if these are equal, all the figures have been found in stage 1.

Hexagons

The number of ways of choosing six pins from nine is:

$$\binom{9}{6} = \frac{9!}{6!\,(9-6)!} = 84.$$

The degenerate figures may be found by a consideration of the $n < 6$ pin figures, bearing in mind the results of lemmas 1 to 3.

Equivalence class					
A3	A4	B4	E4	A5	C5

Number in EC					
4	1	8	4	4	4

Ways of placing remaining pins					
1	6	1	1	1	2

Degenerate patterns

Degenerates					
4	6	8	4	4	8

Pattern A4 gives rise to two different degenerate patterns, one in an equivalence class of four members the other in an equivalence class of two members. There is a total of 34 degenerate figures.

	Pin pattern	Number in EC	Figures possible
A6		4	
B6		4	
C6		8	
D6		2	
E6		4	
F6		8	
G6		2	
H6		8	

I6	4		

The total number of members in the nine equivalence classes is 50. This together with the 34 degenerates makes up the required 84. It has been proved that only 22 different hexagons are possible on the nine-pin geoboard. It is interesting to note that only one non-re-entrant six-pin figure is possible in the nine-pin geoboard. Also, in the case of seven-pin figures there are no non-re-entrants possible on the nine-pin board (see below).

Heptagons

Number of ways of choosing seven pins from nine is $\binom{9}{7} = \dfrac{9!}{7!\,(9-7)!} = 36.$

	Pin pattern	Number in EC	Figures possible
A7		8	
B7		2	
C7		4	

The total number of members in the three equivalence classes found is 14.

	Equivalence class				
A4	C5	C6	E6	H6	
	Number in EC				
1	4	8	4	8	
	Ways of placing remaining pins				
4	1	1	3	1	
	Degenerate patterns				
	Degenerates				
4	4	8	6	8	

It can be seen that the degenerate patterns produced by C6 and H6 belong to the same equivalence class and so the total number of degenerates found should be reduced by eight. Also E6 gives rise to two different degenerate patterns, one with an equivalence class of four and one with an equivalence class of two members, giving a total of six degenerates. Thus the total number of degenerate patterns of seven pins is 22. This result, together with the 14 non-degenerate patterns found gives the required sum of 36 possible patterns. It has thus been proved that there are just five heptagons possible on the nine-pin geoboard.

Octagons

The number of ways of choosing eight pins from nine is:

$$\binom{9}{8} = \frac{9!}{8! \, (9 - 8)!} = 9.$$

The degenerate figures are tabulated below.

A4		Equivalence class B6		A7
		Number in EC		
1		4		8
		Ways of placing remaining pins		
1		2		1
		Degenerate patterns		
• ⊙ •		• • ·		· • •
⊙ · ⊙		⊙ • •		• • •
• ⊙ •		• ⊙ •		• ⊙ •
		Degenerates		
1		8		8

It is seen that the degenerate patterns produced by B6 and A7 belong to the same equivalence class and so the number of degenerate figures found should be reduced by eight. There is a total of nine degenerate figures and so there can be no octagonal figures produced on the nine-pin geoboard. The total number of polygons possible has thus been shown to be 74.

3   *Number Chains*

3.0 Introduction

Number puzzles and the rules of arithmetic provide a source of excellent investigations for primary and secondary school students. Working with sequences of numbers produced from simple arithmetic rules can encourage practice at addition, subtraction and multiplication which is different, and probably a lot more interesting, than traditional methods of learning. Such tasks can also encourage group working as the, often considerable, number of calculations can be shared among different groups in the class. Of course, there is the importance of accurate arithmetic! Again, by working in pairs or larger groups the calculations can be continually checked as each sequence of numbers develops.

The first two articles are closely related, involving the sum and difference of squares. 'Happy Numbers' are probably the best-known and most-used number investigation. It is a nice task to start the process of investigating with primary and lower secondary students. The idea that a number could be 'happy' and another number being 'sad' is quite amusing. The second investigation of 'Sneaky numbers' follows on as a 'what happens if ...?' type question. Perhaps it could be set as a follow up homework type task from a classroom activity on Happy numbers?

The next four articles are closely linked, providing practice at 'times-tables'. The patterns and cycles produced lend themselves to some work with algebra and the notion of proof.

A question often asked by pre-service and in-service teachers is "when can we start to use investigations with students?" 'Bracelets' and the 'two rule and three rule' can be used successfully with students in years 3 and 4 who are familiar with the 2 and 3 times-tables. The tasks provide practice with their tables and the development of a pictorial representation of the cycles. When using this task with primary students, we have found that it is important to have several helpers to continually check the accuracy of the arithmetic. Parents have had great fun with a group of children in after-school and Saturday clubs!

The final article in this Chapter is aimed at upper secondary students and develops number patterns from the mapping $x \rightarrow ax^2 + b$ (mod 100). As the author says in the title 'The fascination of mathematics: Mappings'.

We would argue that the sequences produced in this Chapter from the range of arithmetic rules should provide some fascinating patterns and cycles to explore.

3.1 Michael Mortimer on 'Happy numbers'
Vol. 10, no. 4 (September 1981), pp 27-29; extract

The Greeks were [...] grandiose – amicable numbers, polygonal numbers, perfect numbers and so on. Happy numbers were new to me. David Neal remembered them being played around with at a conference and I've seen them mentioned since [in the ATM's *Points of Departure*] and in what was I think their first published appearance (Jeffery, B. (1978) 'Happy numbers, sad numbers', *Mathematics Teaching* 85, 4-7).

What are they? Well, 44 is happy because we can form a chain:

$$44 \rightarrow 4^2 + 4^2 = 32 \rightarrow 3^2 + 2^2 = 13 \rightarrow 1^2 + 3^2 = 10 \rightarrow 1^2 + 0^2 = 1.$$

The point is that the chain stops at 1. A happy number, then, is one at the start of a number chain which stops at 1. This is a report of an investigation into numbers that are happy. In all likelihood somebody knew all about them a long time ago, and probably still does – Bob Jeffery certainly knows much of what is here, although I read his article near the end of the investigation, and our approaches are rather different. Our investigation uses some computing, and develops a little abstract mathematics along the way. It started with an in-service Primary B.Ed. group on a Wednesday evening. They became involved in the problems rather fast, and by the end of the evening were asking questions we couldn't answer – then! My thanks to them, to David Neal who led the discussion, and to other colleagues who have talked happy numbers with me since then.

Getting started, getting involved

Let's try some more numbers.

$$19 \rightarrow 1^2 + 9^2 = 82 \rightarrow 8^2 + 2^2 = 68 \rightarrow 6^2 + 8^2 = 100,$$
$$7 \rightarrow 49 \rightarrow 97 \rightarrow 130.$$

Now we've got more than two digits, but the process just squares each digit and takes the sum, so:

$$19 \rightarrow 82 \rightarrow 68 \rightarrow 100 \rightarrow 1$$
$$7 \rightarrow 49 \rightarrow 97 \rightarrow 130 \rightarrow 10 \rightarrow 1$$

and all these numbers are happy. Let's try some smaller numbers now:

$$2 \rightarrow 4 \rightarrow 16 \rightarrow 37 \rightarrow 58 \rightarrow 89 \rightarrow 145 \rightarrow 42 \rightarrow 20 \rightarrow 4.$$

Well, 4 occurs twice, so that chain won't ever stop, and every number in it must be sad – we cannot form a chain from that number which will end in 1. And that leads to our first real problem: which numbers are happy?

Mulling and a little insight

The nicest thing would be to find a formula, something like for triangular numbers, but guessing at a formula means searching out a pattern. We need more numbers, so here are a few:

$$3 \rightarrow 9 \rightarrow 81 \rightarrow 65 \rightarrow 61 \rightarrow 37 \rightarrow 58 \rightarrow \mathbf{89} \rightarrow 145 \rightarrow 42 \rightarrow 20 \rightarrow 4,$$
$$6 \rightarrow 36 \rightarrow 45 \rightarrow 41 \rightarrow \boxed{17} \rightarrow 50 \rightarrow 25 \rightarrow 29 \rightarrow 85 \rightarrow \mathbf{89},$$
$$14 \rightarrow \boxed{17}.$$

They are all sad, and more than that, parts of the chains are the same. Then it becomes clear, perhaps, that the order of the digits *doesn't matter*; 25 is sad, so 52 is as well, and even 205! Also, we can extend chains backwards. For example,

$89 = 5^2 + 8^2, 58 = 3^2 + 7^2$ and $85 = 2^2 + 9^2$, so we get a tree pattern.

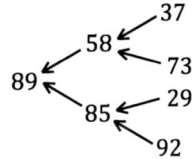

Keep going

We might profitably limit our sights at this stage, and ask only for all happy numbers below, say, 100. Using a number square to keep track helps a lot, and with application we can construct a full tree pattern of happy numbers below 100:

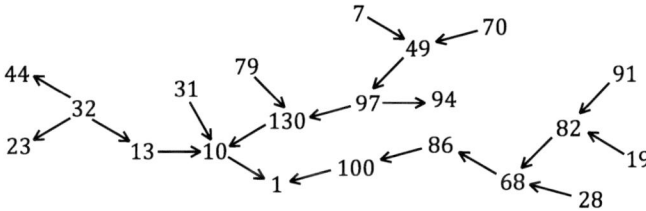

(Watch out for 130, it's easily dismissed as too big.)

Working backwards up the tree uses essentially two ideas:

a) the order of the digits doesn't matter - and neither do any extra zeros, so since 19 is happy, so are 91, 109, 910, etc.
b) a number in the chain has a predecessor if it is a sum of squares.

Of course, by taking enough 1's, every number is a sum of squares, but here we're really interested in two-digit numbers, so we could get involved in questions of when a number is the sum of two, three, or even four squares. Perhaps however, it would be best to keep to the problem in hand.

Looking back and checking

Unfortunately we still have no real answer to our question: Which numbers are happy? However, looking back at the sad numbers shows a rather remarkable pattern. Not only do number chains for sad numbers overlap, but in every case the chain from a sad number seems to attach itself to the closed chain [alongside].

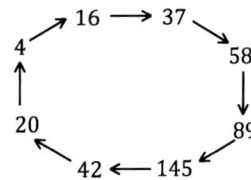

If that were always true, then rather than a split between happy numbers and sad numbers (i.e. chains going to 1 or not going to 1), we would differentiate between those with chains going to 1 and those with chains going to 4. This would give us the next best thing to an explicit formula for happy numbers, namely a potential computer program which for any particular number would tell us in a finite amount of time whether the number is happy or sad, i.e. whether the set of happy numbers is recursive.

At this point, Michael Mortimer gives a BASIC program to check that all the positive integers to 1000 are happy, together with a little theory. He draws attention to the fact that the links of any chain do not descend necessarily but where they ascend it is short-lived. He goes on to suggest ways to develop the investigation and to draw his conclusions, which is where we pick it up. Perhaps teachers and pupils could write some appropriate code (Eds.)?

It might be tempting to look at sums of cubes or of fourth powers ... My own preference has been to look at our original function f, but using different number bases. To whet your appetite, every number is happy in base 2 and in base 4. (*The Editors have tried the base 4 investigation below; a colour version is on the front cover*). Investigating happy numbers in bases other than 10 presents some rather nice problems – I commend them to you.

We started with a fairly simple idea for forming number chains. My claim for this particular one (among the many) is that not only does it lead to some intriguing changes of perspective – in working back up number chains, in switching from happy to sad numbers, etc., but it brings in the computer in a realistic and useful way as a means to an end, as well as allowing some fairly abstract, but nonetheless accessible mathematics. Quite where any investigation should stop in the school situation depends on interest, tenacity, and level of attainment, as indeed does the direction in which it might go. The article by Jeffery is worth reading for its differences as well as its similarities to parts of this article. For this investigation I think there is material of value for teachers of children in the junior school right through to the sixth form. Try it and see what happens!

3.2 'Sneaky numbers: A source of investigations' by Ian Brown
Vol. 20, no. 3 (May 1991), pp 12-13

Teachers looking for some interesting investigations may like to exploit the mathematics surrounding the so-called 'sneaky' numbers. Although I had heard of 'happy' numbers, I had not met sneaky ones till I found them mentioned by Kirkby in one of his useful booklets. The ideas suggested in this article should provide valuable motivation for the first school child to develop basic skills with multiplication, subtraction etc. and an opportunity to acquire language skills by describing numbers and assigning them human attributes. Older children will enjoy producing their own 'tree' or 'flower' diagrams (see solutions to 7 and 8 below) classifying all two digit numbers, while sixth formers may enjoy writing a computer program to check these results and generalising the idea beyond two digit numbers. What is a Sneaky Number? Consider a two digit number like 95. Call the digits a and b. Take the positive difference of the squares of the digit. If you now have a single digit number then STOP. If you still have a two digit number then repeat the process. A number is 'sneaky' if you finally reach ZERO.

Example 1: Is 95 sneaky?

$$95 \rightarrow 9^2 - 5^2 = 81 - 25 = 56,$$
$$56 \rightarrow 6^2 - 5^2 = 11,$$
$$11 \rightarrow 1^2 - 1^2 = 0.$$

We have reached 0, so 95 is sneaky.

Example 2: Is 96 sneaky?

$$96 \rightarrow 9^2 - 6^2 = 45 \rightarrow 5^2 - 4^2 = 9.$$

We have reached 9, not 0, so 96 is not sneaky.

Example 3: Is 17 sneaky?

$$17 \rightarrow 7^2 - 1^2 = 48,$$
$$48 \rightarrow 8^2 - 4^2 = 48,$$

We are going round in circles! We shall never reach 0, so 17 is not sneaky.

Investigations

1. Find as many sneaky numbers as you can. (Test only the two digit numbers 10 to 99.)

2. Which numbers get to 0 in one step? What shall we call them? 'Boringly sneaky'?

3. Which number takes the most steps to get to 0? Shall we call this the 'sneakiest' number?

4. Can you find more numbers which end up stuck on one number? (like 17 got stuck on 48 in Example 3) 'Sticky' numbers?

5. Are there any other ways of getting stuck? Try 37. What would you call 37? A 'loopy' number?

6. Which numbers always decrease on their way to zero? You might call these 'downhill' numbers. (Older mathematicians might speak of a 'monotonic' or strictly decreasing sequence). How many non-boring downhill sneaky numbers are there?

7. Draw a diagram to show how all the sneaky numbers make their way to 0.

8. Draw similar tree diagrams to show the non-sneaky numbers and how they either reach a non-zero single digit or get stuck.

9. Which is the most popular single digit to reach (other than zero)? Are any digits never reached?

10. Show on a bar chart or in a table how many of the 90 two-digit numbers reach each of the digits 0 to 9 and how many are sticky or loopy?

Answers

1. There are 90 two-digit numbers; 22 of them are sneaky.

2. The multiples of 11 are 'boring' because, for example: $44 \rightarrow 4^2 - 4^2 = 0$.

3. 70 is the sneakiest number because it takes the longest to get to 0 (4 steps).

4. 17, 71 and 84 get stuck on 48 after one step; 89 and 98 take 2 steps.

5. 37, 40, 53, 61 and 73 get stuck in a loop $16 \rightarrow 35 \rightarrow 16 \rightarrow 35$ ...

6. The following numbers are strictly decreasing as they approach zero: 11, 22, 33, 44, 47, 55, 56, 59, 65, 66, 74, 77, 83, 88, 92, 94, 95, 99. If we delete the nine multiples of 11 (boring) from this list we have 9 non-boring downhill sneaky numbers.

7.

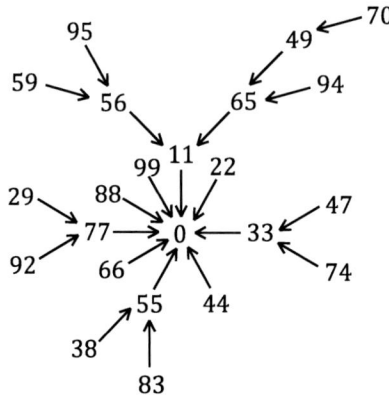

8.

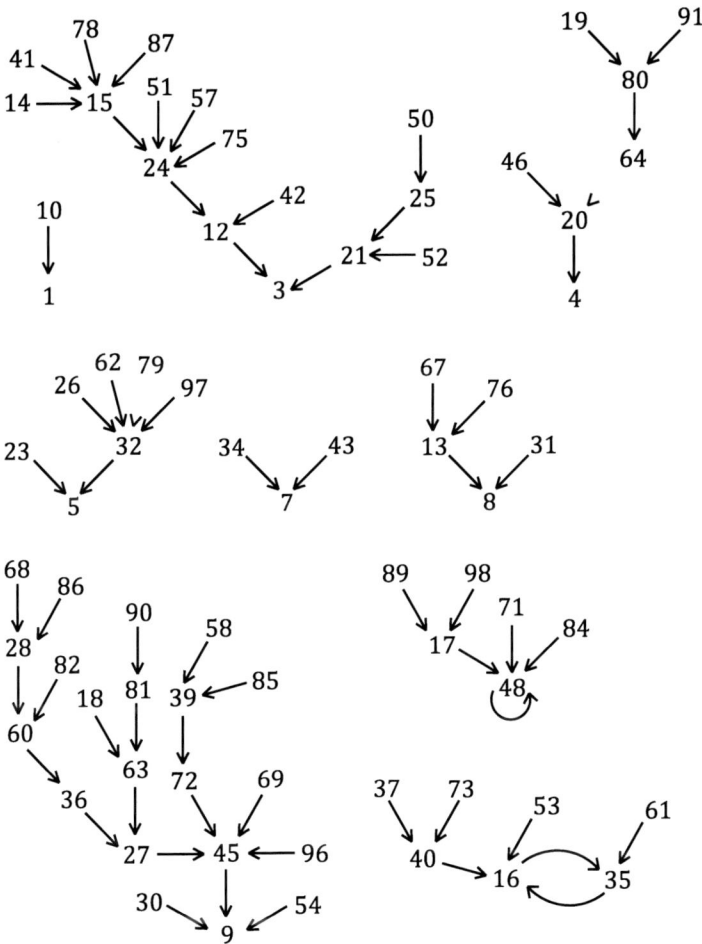

9. The most popular digit to reach (other than zero) is 9. The numbers 2 and 6 are never reached.

10.

Final digit	0	1	2	3	4	5	6	7	8	9	None (loop)
Frequency	22	1	0	15	6	6	0	2	4	21	13

References

Kirkby, D. (1986) *Square Numbers 2, Investigation Bank Books 4*, Eigen Publications.

Wells, D. (1986) *The Penguin Dictionary of Curious and Interesting Numbers*, Penguin Books.

3.3 John Berry's 'Bracelets'
Vol. 43, no. 5 (November 2014), pp 31-34

The article by Tony Orton (*see p. 117*) reminds me of a very investigation on numbers that I first came across in the early 1980s whilst working with David Burghes and the Spode Group. The basic problem is this:

1. Choose any integer between 1 and 18.
2. Identify the units and tens digits in your integer and form a new number by multiplying the units digit by 2 and adding the tens digit. For example, if you choose 16 then the new number is $6 \times 2 + 1 = 13$, if you choose 7 then the new number is $7 \times 2 + 0 = 14$.
3. Repeat step 2 for the new number.
4. Keep going and see what happens. This is called the 2-Rule. The outcome is a cycle of the integers 1 to 18 shown in Figure 1.
5.

Fig. 1: The cycle of 18 integers formed by the 2-Rule

Several questions spring to mind:

• What happens if we start with an integer greater than 18?
• What happens if we start with a three-digit integer?
• What happens if we multiply by 3 instead of 2?

For the 2-Rule we find that all the integers greater than 19 and less than 99, except the multiples of 19, join the cycle after either one 'step' or two 'steps'. For example:

$91 \rightarrow 1 \times 2 + 9 = 11$ (in the cycle after one step),
$96 \rightarrow 6 \times 2 + 9 = 21 \rightarrow 1 \times 2 + 2 = 4$ (in the cycle after two steps).

Also, the integer 19 maps to itself, i.e. $19 \rightarrow 9 \times 2 + 1 = 19$, and all the multiples of 19 are mapped to 19. So in a sense 19 is a 'special number' for the 2-Rule. Figure 2 shows the outcome for several starting integers greater than 19. You can perhaps see why I called this investigation 'Bracelets'? The initial cycle is our chain in the bracelet and the numbers 'hanging' from the chain are the charms!

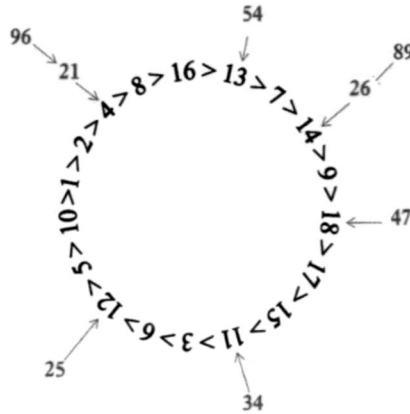

Fig. 2: A bracelet for the 2-Rule

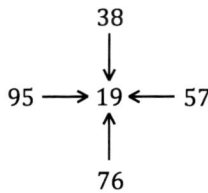

Fig. 3: Mappings for the 'special' number 19

A three-digit number is just as easy to deal with. Consider the integer 345. The units digit is 5 and the 'number of tens' is 34 – this becomes the tens digit in the 2-Rule. $345 \rightarrow 5 \times 2 + 34 = 44 \rightarrow 4 \times 2 + 4 = 12$ (in the cycle after two steps). During the 1990s the '2-Rule' developed into an investigation for Year 6–8 pupils as a half-day mathematics workshop with the aim of developing pupils' investigative skills and using the skill of asking "what happens if ...?" questions.

The investigation has four phases:

First phase: 2-Rule as a whole-class activity.

Second phase: 3-Rule in pair-groups and a whole-class feedback.

The 3-Rule has a similar outcome to the 2-Rule shown in Figure 4. There is one cycle of the integers 1–28 with 29 being the 'special' number in that 29 maps to itself and multiples of 29 map to 29.

Third phase: After exploring the 2-Rule and the 3- Rule, there is an opportunity for reflection and prediction of what might happen for the other rules. Table 1 shows how pupils are led into making sensible predictions. We start with the rows for the 2-Rule and 3-Rule completed and ask for what might happen for the rest.

Fig. 4: The cycle of 28 integers formed by the 3-Rule

Making conjectures on their own and exploring them is an important part of 'thinking like a mathematician' and this investigation provides such an opportunity for primary and lower secondary pupils. It is probably a skill that is rarely developed in our rather narrow, test-focussed mathematics curriculum!

Fourth phase: Now share out the exploration of the 4-, 5-, 6-, 7-, and 8-Rules to small groups in the class.

Rule	Cycle of numbers	Special number
2	**1 to 18**	**19**
3	**1 to 28**	**29**
4	1 to 38	39
5	1 to 48	49
6	1 to 58	59
7	1 to 68	69
8	1 to 78	79
9	1 to 88	89

Table 1: Pupil predictions of the outcomes for rules
other than the 2-Rule and 3-Rule

Those pupils doing the 4-Rule are usually the first to find out that the predictions in Table 1 do not work. They soon find a small cycle of just six numbers. This result usually creates a buzz in the classroom as the other groups wonder if their predictions from Table 1 will work or not. Figures 5–9 show the results for the 4-, 5-, 6-, 7-, and 8-Rules. Only one rule, the 6-Rule, agrees with the prediction. Furthermore, the 4-Rule, 6-Rule and 7-Rule each have more than one 'special' number. The 9-Rule has two large cycles of 44 integers and I feel that it is rather cruel to encourage a group of pupils to get through these cycles!

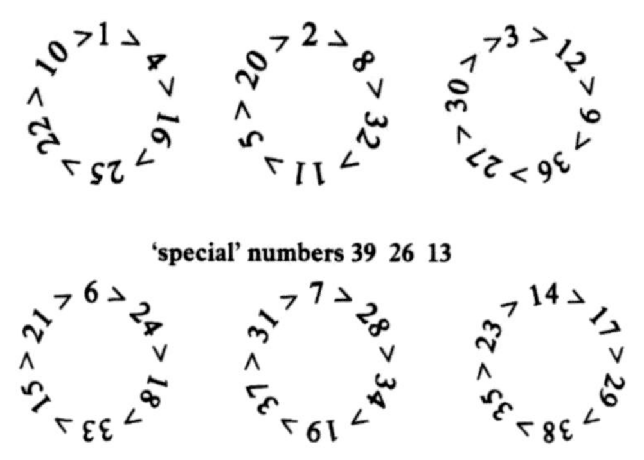

'special' numbers 39 26 13

Fig. 5: The cycles of integers and 'special' numbers formed by the 4-Rule

An alternative fourth phase: An interesting alternative is to omit the 8-Rule from the feedback to the class from the pupil groups. I encourage the 8-Rule pupils to keep their outcomes secret. As feedback to the outcomes, Table 1 can be improved to Table 2 below. A whole-class discussion usually results in the idea that if the 'special number' is a prime number then there will be more than one cycle. Now the 8-Rule group can demonstrate that this prediction fails and, although 79 is the 'special' number and is prime, there is more than one cycle.

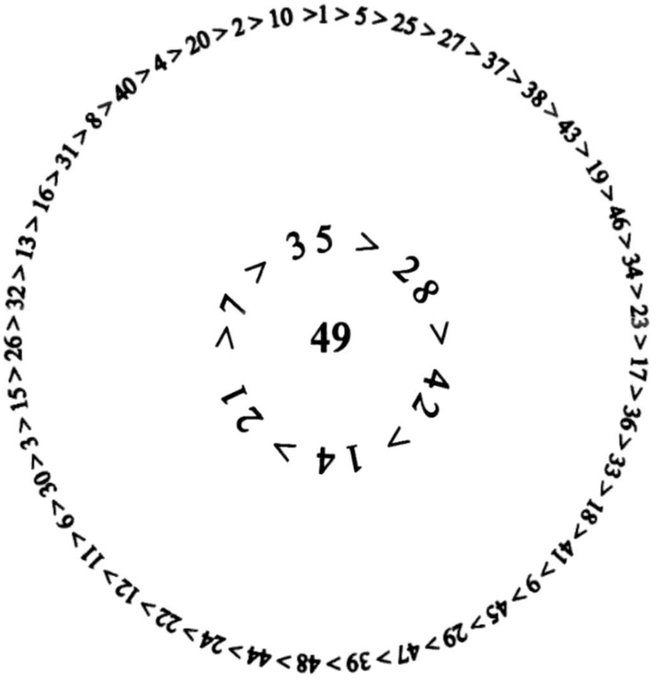

Fig. 6: The cycles of integers and 'special' number formed by the 5-Rule

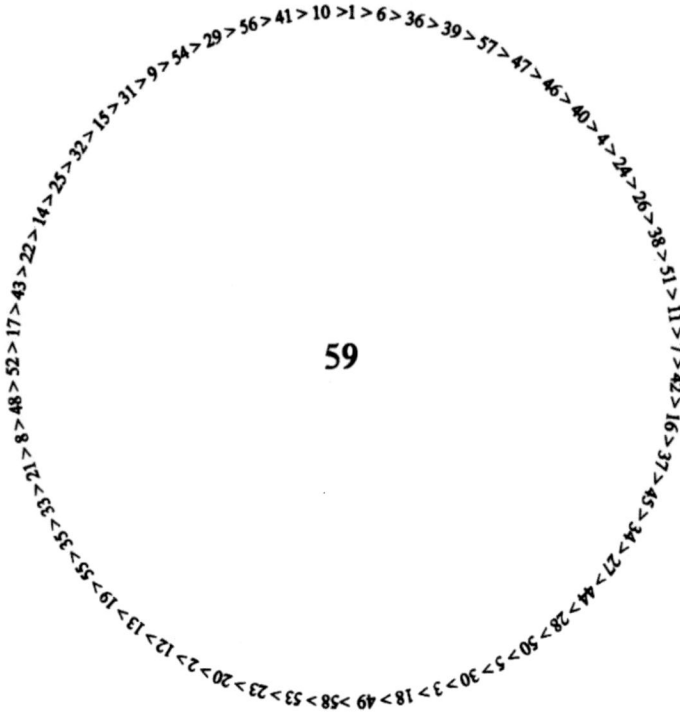

Fig. 7: The cycles of integers and 'special' number formed by the 6-Rule

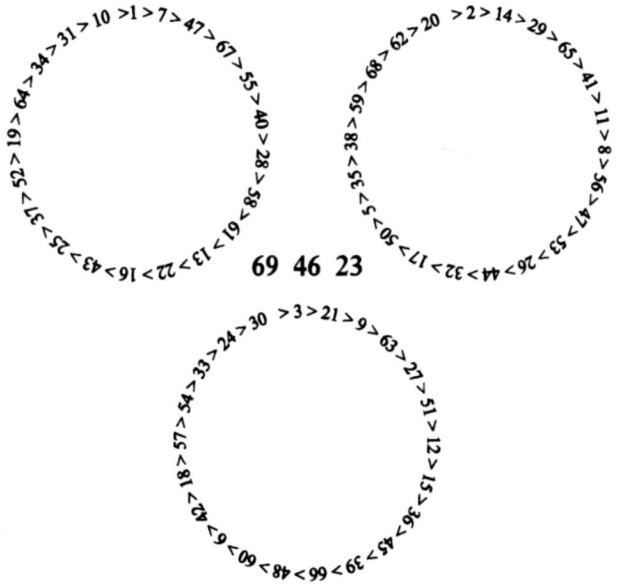

Fig. 8: The cycles of integers and 'special' number formed by the 7-Rule

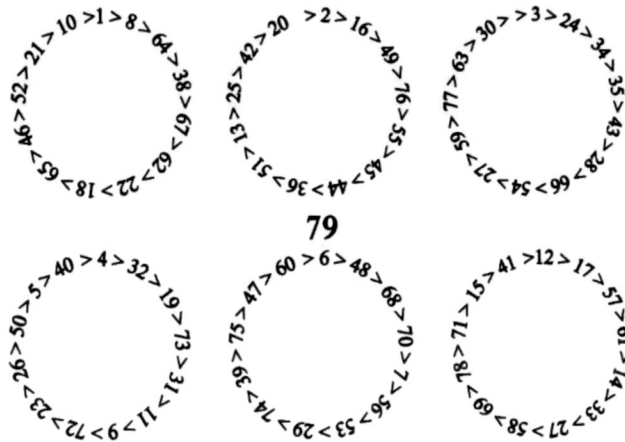

Fig. 9: The cycles of integers and 'special' number formed by the 8-Rule

Rule	Number of cycles of numbers	Special numbers
2	1	19
3	1	29
4	more than 1	13, 26, 39
5	more than 1	49
6	1	59
7	more than 1	23, 46, 69
8	?	79

Table 2: Pupil outcomes for rules up to the 7-Rule

This investigation is very pupil-active and has proved very popular with primary and lower secondary pupils over the years. It provides the opportunity to:

- make predictions and try to verify them
- practise the times-tables (no calculators allowed!)
- work cooperatively in groups; pupils can start with different starting numbers in their group and see when they 'catch up' with another group member
- ask "what happens if ...?" questions and follow them up
- make posters of their outcomes to show to the other groups.

An algebraic approach to the Bracelets Investigation

For a few pupils, in lower secondary school, there is the opportunity to explore some algebra of the cycles and the special numbers.

Finding the special numbers algebraically

The special numbers in each rule are the numbers which map to themselves and are summarized in Table 3. Can we find them algebraically? Consider the two-digit number ab where a and b are integers with $1 \le a \le 9$ and $0 \le b \le 9$.

To work with ab algebraically, we write it as $10a + b$.

Suppose that first we consider the 2-Rule so that $ab \rightarrow 2b + a$.

Then if ab is a 'special' number that maps to itself, we have

$$2b + a = 10a + b \Rightarrow b = 9a.$$

The only solution for this equation in the intervals $1 \le a \le 9$ and $0 \le b \le 9$ is $a = 1$ and $b = 9$. This proves that for the 2-Rule, 19 is the 'special' number.

The proofs for the 3-, 5-, 6-, 8- and 9-Rules are very similar with just one solution in each case. It is the algebra for the 4-Rule and 7-Rule that interest the pupils most.

For the 4-Rule we have $4b + a = 10a + b \Rightarrow 3b = 9a \Rightarrow b = 3a$. There are three solutions: $a = 1$ and $b = 3$; $a = 2$ and $b = 6$; $a = 3$ and $b = 9$ giving the 'special' numbers 13, 26 and 39 for the 4-Rule.

For the 7-Rule we have $7b + a = 10a + b \Rightarrow 6b = 9a \Rightarrow 2b = 3a$. There are three solutions: $a = 2$ and $b = 3$; $a = 4$ and $b = 6$; $a = 6$ and $b = 9$ giving the 'special' numbers 23, 46 and 69 for the 7-Rule.

Rule	Special number
2	19
3	29
4	13, 26, 39
5	49
6	59
7	23, 46, 69, 79
8	89
9	

Table 3: The special numbers for each rule

Do the numbers map to a smaller number?

There is a nice proposition that we can make for each of the rules.

For the whole-class activity exploring the 2-Rule the pupils show that all integers in the range 1–18 produce a cycle, and all integers greater than 18 either map to a number in the cycle (after one or more steps) or map to 19 (for all the multiples of 19).

Pupils often ask: "Is it true that all integers greater than 19, which are not multiples of 19, will map to the cycle?" My response is "Yes, because the numbers always get smaller after each mapping until they 'hit' the cycle."

An algebraic proof is quite nice for those pupils who have sufficient algebra. Suppose that there is an integer AB with $A \geq 2$, which increases after applying the 2-Rule. In algebraic form the proposition is:

find values of A and B so that $2B + A > 10A + B \Rightarrow B > 9A$.

Since $A \geq 2$ then $B > 9A \Rightarrow B > 9$ but this is not possible because for a two-digit number $0 \leq B \leq 9$.

Thus there is no integer greater than 19 which maps to a larger number, i.e. after applying the 2-Rule each integer greater than 19 will get smaller until it 'hits' the cycle or the special number 19. A nice example of proof by contradiction! Similar proofs occur for the other rules.

Reference

Orton, T. 2012 'The Fascination of Mathematics: Mappings', *Mathematics in School*, 41, 5, pp. 35–44.

3.4 'Times tables tests from iterations' by Gordon Haigh
Vol. 19, no. 2 (March 1990), pp 7-8

Pick a number, say 27: separate the rest from the units and regard as two numbers, here 2 and 7. Create a new number by taking 2 (the rest) + 3 (units), here $2(2) + 3(7) = 25$, so $27 \rightarrow 25$. Continuing this gives a loop:

$$27 \rightarrow 25 \rightarrow 19 \rightarrow 29 \rightarrow 31 \rightarrow 9 \rightarrow 27 \dots$$

Rule

$$
\begin{array}{ccc}
 & 2 \quad 7 & \\
\times 2 & \downarrow \quad \downarrow & \times 3 \\
 & 4 + 21 & = 25
\end{array}
$$

Doing this for the first few numbers gives this partition:

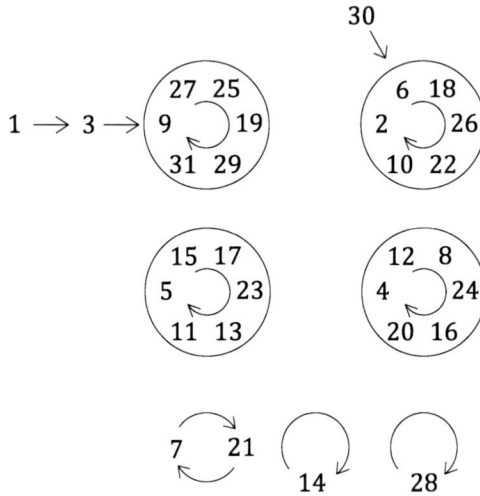

Things to notice about this are the numbers in the 4 times table are isolated. Also evens go to evens and 7 times table goes to 7 times table. This turns out to be true for higher numbers as well, so the rule can be used as a test for these tables. For example,

116? $2(11) + 3(6) = 40; 2(4) + 3(0) = 8$, so 116 is in the 4 times table.

1001? $2(100) + 3(1) = 203; 2(20) + 3(3) = 49; 2(4) + 3(9) = 35;$
       $2(3) + 3(5) = 21$, so 1001 is in the 7 times table.

The clue to inventing them is given by the numbers that map to themselves: take a number with more than one digit, split it into two parts, take linear multiples of the parts to recreate the original number and the chances are you will have a test for that number's table and in fact the table of any factor of that number. e.g. take 39; 1 (the rest) + 4 (units) = 39 so trying this out on the low numbers gives:

Rule

$$\begin{array}{cc} 2 & 7 \\ \times2 \;\downarrow & \downarrow \;\times5 \\ 4 + 35 & = 39 \end{array}$$

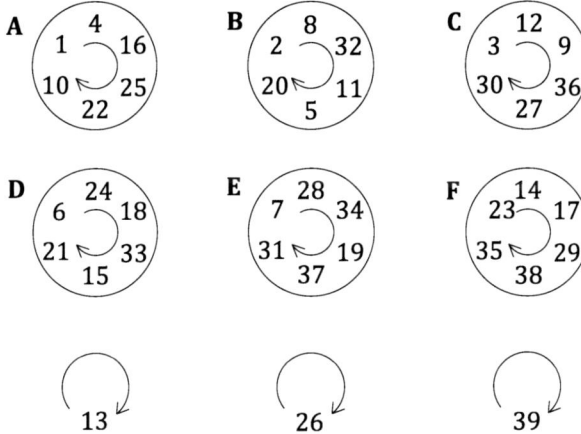

This shows the numbers nicely partitioned into $3n$ numbers (set **C** and set **D**), $3n + 1$ numbers (sets **A** and **E**) $3n + 2$ numbers (sets **B** and **F**) and the 13 times table numbers. So this is a test for the 3 and 13 times table. As both these diagrams hint at, a manipulation of a number so that it gives another number in the same table will often work also, e.g. 21? $1(2) + 5(1) = 7$, so 1 (the rest) + 5 (units) is a test for the 7 times table. There seems to be work here for secondary school investigations and perhaps primary school number skills – it's a bit like those 'happy numbers' investigations but with more connections to other parts of the syllabus.

Rule

$$\begin{array}{cc} 2 & 7 \\ \downarrow & \downarrow \;\times4 \\ 2 + 28 & = 30 \end{array}$$

The reasons why they work are discussed by D. P. Eperson in *Mathematics in School* 16, 5 (November 87), where he considers two 7 times table tests and argues something like this: given a number is in the 7 times table, then altering it by a multiple of 7 will leave it still in the 7 times table. Denoting such a number by $10a + b$ thereby separating the rest from the units, the rest = a the units = b, we can therefore add on 49 bs and still be in the table so

$$10a + b \rightarrow 10a + b + 49b = 10a + 50b = 10(a + 5b).$$

Now 10 isn't in the 7 times table so $a + 5b$ must be, and this is the test we use. The question may arise as to how to predict the length of the cycles. I cannot work this out but have a clue that it is linked to the repetend length of the reciprocal of the number whose table is being tested. (Which suggests it's to do with primitive roots.) Is there a rule of this type that never cycles? Finally here is the rich 2 (the rest) + 5 (units) which is a test for the 2, 3, 4, 6, 8, 12, and 24 times tables!

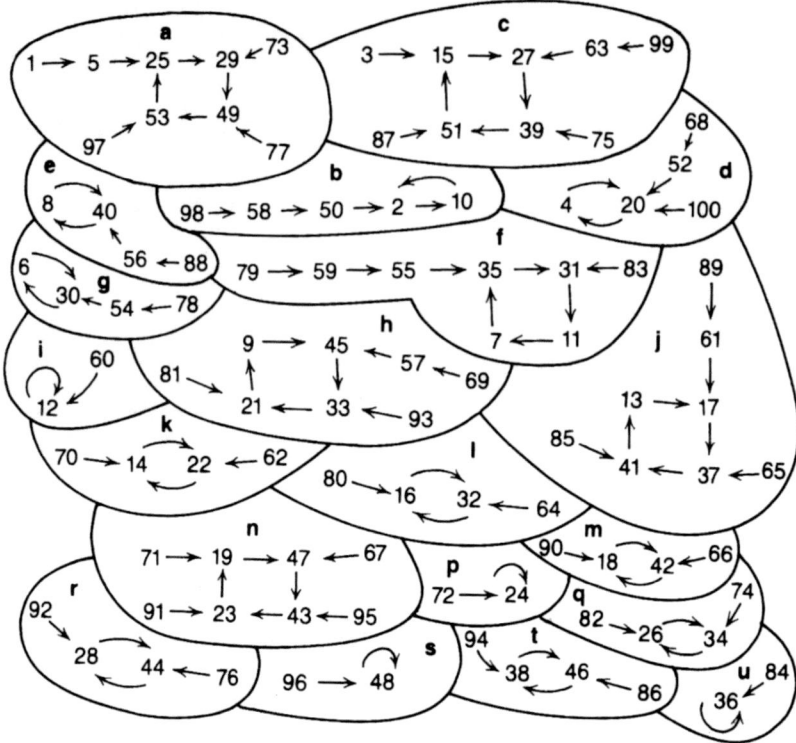

2 times table	sets **b, d, e, g, i, k, l, m, p, q, r, s, t, u**
3 times table	sets **c, g, h, i, m, p, s, u**
4 times table	sets **d, e, i, l, p, r, s, u**
6 times table	sets **g, i, m, p, s, u**
8 times table	sets **e, l, p, s**
12 times table	sets **i, p, s, u**
24 times table	sets **p, s**
$4n + 1$ numbers	sets **a, h, j**
$4n + 3$ numbers	sets **f, n**
$8n + 2$ numbers	sets **b, m, q**
$8n + 6$ numbers	sets **k, t**
$12n + 3$ numbers	set **c**
$12n + 9$ numbers	set **h**
$16n + 4$ numbers	set **d**
$16n + 8$ numbers	set **e**
$16n + 12$ numbers	set **r**
$24n + 6$ numbers	set **g**
$24n + 12$ numbers	set **u**
$24n + 18$ numbers	set **m**

3.5 'Much more than multiplying by 5' by Andrejs Dunkels
Vol. 20, no. 3 (May 1991), pp 9-11

For many years I have been using number sequences with students of different ages, from primary children to pre-service and in-service teachers. Number sequences are useful as substitutes, or complements, of traditional drill practice. With the traditional type of isolated, 'naked' exercises the learner does not experience any connection between successive computations, even if the author has designed the unit under consideration in a particular way. Working with sequences is different. Computations are not performed in isolation, there are connections and inter-relations to be explored and discovered, and the learner gets used to being observant about what is happening, not least as related to previous calculations.

In 1987, I used for the first time one of the sequences mentioned in [Gordon Haigh's] article. The idea came from Kennedy et al (see reference). I needed a suitable activity for reviewing the multiplication tables up to 10×10. I wanted an activity that would work as a diagnostic tool for the pupils and that differed from exercises of traditional style. This is how it happened.

Let us take 18 as our starting number, I said at the beginning of the lesson. And then we will successively create new numbers. I wrote 18 on the board. To me working with sequences is part of 'lab work with numbers'. I use this phrase to indicate the working method. Usually 'lab work' is associated with practicals, like weighing or measuring physical objects, or pouring liquids into various cups. However, one can work in very much the same way with numbers as objects, with emphasis on exploration and discovery.

For this particular class today's sequence was not the first encounter with lab work with numbers.

> There is a rule, I went on, for the next number: Delete – and I do mean delete, or erase, not subtract – so, delete the units digit. Then, to what's left add 5 times the deleted number. In our case this means that we first delete 8, which leaves us with 1, and then to this 1 we add 5×8. On the board I added an arrow and 41, so this is what we now had on the board.

$$18 \rightarrow 41$$

I did not present the rule in writing, only orally. Neither did I write down any intermediate steps. I wanted the pupils to try to keep all this in their heads. So our new number is 41, I continued. We will now carry on and apply the same rule to the new number in order to get yet another new number. I asked one of the pupils to explain the first step of the rule, another to explain the second step, and a third to state the result. In summary: delete the units digit, that is delete 1, leaving 4; to what's left add 5 times the deleted number, that is $4 + 5 \times 1$, which gives 9. I recorded that on the board. I knew beforehand that we would get a one-digit number here when starting with 18, and I wanted that situation to occur so that we could sort out how to apply the rule to one-digit numbers. If we delete the units digit from 9 there is nothing left. This situation is, of course, handled using zero, and so the next number after 9 is $0 + 5 \times 9 = 45$. I recorded the new number on the board. Now the time had come to ask the pupils to work out the next number in the head without saying anything, and when I felt the majority was ready one of the pupils was asked to give a suggestion, which I recorded on the chalk board. Another pupil was asked to explain how the result was obtained. This is what we now had on the board.

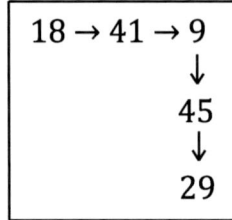

$$18 \to 41 \to 9$$
$$\downarrow$$
$$45$$
$$\downarrow$$
$$29$$

Alright, I said, now you can continue on your own. A natural question arose, although the pupils were familiar with exploratory activities:

For how long are we to continue?
Until you discover something ...

And what are we supposed to discover?
How should I know! It is in the nature of discovery that you don't know at the outset what you will discover.

Everybody got started, and there were occasional comments about what to expect (I know, we're going to end up with 1 / We'll get 18 back after a while / There's going to be a pattern, that's what it's going to be). And the pupils worked on. At this stage those readers who have not already tried this thrilling sequence are encouraged, before reading further, to have a go at the same task as the pupils were given.

I had not prepared this lesson more than discovering that 18 would indeed reappear, making the sequence a loop, and that this takes some time. Following Kennedy, I had also noted that the loop does not contain all numbers from the smallest, 1, to the largest, 48. I had also found all the missing numbers to be from the sevens table, and discovered that they form a loop of their own with one of them as the starting number and the same rule applied. That was all I knew at the time I started this lesson. The idea about smallest and largest gradually spread through the class, partly via the jungle telegraph. I tried to reveal as little as possible.

Before they got back to 18, some pupils said that they were certain there would never be an end to this sequence. There would be new numbers all the time, they thought. I think that it is necessary to have activities that last for quite a while. I fear that otherwise the Instant Coffee Syndrome (the belief that in mathematics everything is solvable within the time it takes to make a cup of instant coffee) will spread and do much damage to the way people view mathematics and its use for problem solving.

The pupils thoroughly enjoyed finding the missing numbers and discovering that all are from the sevens table. In Sweden, parents are encouraged to come to school and participate in their children's work. To this lesson Jon's mother had come. It was therefore particularly exciting that he made the discovery of the day – "the numbers double", Jon said to me. His mother who had been going around in the classroom watching all the pupils work was at the moment right behind him. I asked him to explain more, for I honestly did not understand what he meant. "Look", he continued, "here we have 15 and here is 30. Again, here is 8, 16, and 32, doubles." "Hmm, I hadn't seen that", I said, and Jon added, "It's like jumping two steps at a time backwards and doubling."

At that point I thought it would be appropriate to let Jon tell the class about his discovery, but I asked him to do the telling rather vaguely so that the rest of the class would get a chance of doing some exploration. This sparked an interest and many of Jon's classmates rediscovered Jon's discovery – with a lot of discussion and exchange of views, ideas, and comments. Some of the pupils checked for doubling in the 7-table-loop.

Now there was time for some summing up. It is important, I feel, to rewrite one's rough notes in some form, or forms, so as to get an overview of the situation. Some pupils had counted how many different numbers we had in the big loop, and we decided to start with 1 and write half of the numbers from left to right and the other half going back to 1 from right to left (see below). This rewriting, of course, took longer for some pupils, and Helena who was ready among the first asked, "What happens if I take 50 as the starting number?" "I don't know off the cuff", I replied, "but I like your question, so why don't you try!"

I had hardly turned my back to Helena when she called me back, and in an excited voice said,
Look! 50 is like 1!
Like 1?! How is that?
From 1 you get to 5, and that's what you do from 50 as well.
Indeed.
And I know what's going to happen with 51.
You do?
Yes, 51 will be like 2.

And she tried her hypothesis and was very pleased indeed when it turned out to check. "I know that 52 will be like 3", she went on, checked and predicted what would happen with larger numbers. She got happier and happier each time she verified her hypothesis.

As other pupils got ready with their rewriting of the loop they too started investigating numbers from 50 and up. This figure shows the rewritten record of one of the pupils.

By the way, when Helena reached 56 she immediately said, "I know. This will lead me to the sevens loop." Of course she was right.

I cannot tell why and how we managed to avoid trying 49 throughout the whole lesson. But we did. It was not until a week later that I discovered that 49 is very special for the rule under consideration. Those readers who have not yet tried 49 as the starting number should do so before reading further. I discussed the activity with a group of in-service teachers, and we wanted to go more deeply into this rule. The question, 'Why the sevens table?' had been asked and, in order to find an answer, we had decided to try some algebra, introducing the following notation: n = next number, s = starting number, u = units digit of s, t = what's left when u has been deleted from s (or number of tens in s). Then we have

$$n = t + 5u \quad (1)$$
$$s = 10t + u \quad (2).$$

In order to see what is going on the ideal situation would be to eliminate t and u from (1) and (2) and express n in terms of s alone. Some trials and a little inspection indicates that this might not be possible, however. The second best would be to eliminate one of t and u, u say, and express n in terms of s and t. So we keep (1) unchanged and multiply (2) by 5 which gives

$$n = t + 5u \quad (1)$$
$$5s = 50t + 5u \quad (3).$$

Now we subtract (3) from (1) and obtain

$$n - 5s = t - 50t + 5u - 5u \quad (4).$$

Now watch out, for the next step is important and revealing, it involves $t - 50t$ which is $-49t$, and so 49 pops up and (4) gives $n = 5s - 49t$ (5).

Having determined (5), it is easy for the experienced user of mathematics to conclude that our rule partitions the set of all positive integers into three disjoint classes, one containing all numbers without the factor 7, one containing all numbers with exactly one factor 7, and one containing all numbers with two or more factors 7. Details are left to the reader.

The in-service teachers and I found it rather entertaining to check 98, which is 2 × 49, as a starting value. 392 might also be tried.

Incidentally, during this meeting we had actually discovered that if we have the big loop recorded, as in the pupil's drawing above, then the numbers 1 and 48 appear at the extremes, and the sum of these numbers is 49. The magic 49 again. With the in-service teachers I also went further with the doubling phenomenon. What happens when doubling does not occur? If we jump two steps back from 32 we do not reach the double 64 but 15, if we jump in the same way from 44 we do not end up in 88 but 39, etc. We made a table:

Number	Double	Actual next number
32	64	15
44	88	39
48	96	47
40	80	31

At this point the reader might find it interesting to do some exploring of this before reading on. The idea is to see what happens if 'the double' is chosen as the starting value and applied the same rule. For us this led to thoughts similar to those of Helena. Another way was to see what, if anything, the two right-most columns in our table have in common. The details of this, including the algebra, are left to the reader.

There is more to explore here. I had pupils from another class write the loop in a rectangular shape as near a square shape as is possible with, say, 1 in the left upper-most corner. It was not surprising to find that 48 occupied the diagonally opposite corner to 1, but it was indeed surprising that there were so many more 49-sums to be discovered. The proof of this turns out to be rather demanding on the side of experience with algebraic expressions, and I have treated the proof only at some in-service meetings with secondary school teachers. I have not tried it with secondary students, but as far as I can see this sequence provides some nice concrete algebra adventures suitable for secondary education. I leave the details of this to the reader and his or her students.

A natural query is of course if the multiplication by 5 in the rule is essential or if other multipliers can be used. As I see it a teacher can take two different stances in this situation:

a) I have to try all this myself first before I bring it to my classes. (Seven years later the teacher will enjoy doing it with the pupils.)

b) I will tell my pupils that I don't know all of this and that we'll have to do discoveries together. (The pupils will benefit from experiencing situations where the teacher really does not know all the answers. And the teacher will have to cope with the fear that this at first might cause.)

After I had used, for some time, the rule with 5 as the multiplier I started believing that I knew all there is to know about this particular rule. It was therefore pleasing, and somewhat surprising, when a group of teachers at an in-service workshop made a new discovery after having rewritten the big loop. Since that loop contained 42 numbers they had chosen to follow standard writing conventions and write the numbers in 6 rows with 7 numbers to the row. The group now associated to magic squares, where, as is well known, one adds rows and columns. The reader might, perhaps, at this point wish to spend some minutes on the idea of this group before reading further.

1	→	5	→	25	→	27	→	37	→	38	→	43	→
19	→	46	→	34	→	23	→	17	→	36	→	33	→
18	→	41	→	9	→	45	→	29	→	47	→	39	→
48	→	44	→	24	→	22	→	12	→	11	→	6	→
30	→	3	→	15	→	26	→	32	→	13	→	16	→
31	→	8	→	40	→	4	→	20	→	2	→	10	→

The row sums turned out to be all different, and the group did not see any pattern, and the surprise came with the columns which all added up to 147. They called me, and I was at first as astonished as they were. Eventually we managed to relate this to previous discoveries. Note that we have $147 = 3 \times 49$. That magic 49 again! Now each one of the first half of the 42 numbers in the loop has a counterpart, a 49-complement, on the third row down. Thus two such numbers add up to 49. Therefore this discovery is just a nice straightforward application of the previously discovered 49-sum property of our loop.

Lab work with numbers has given me and my students lots of enjoyable moments as well as valuable mathematical experiences. In particular, this class of sequences has catered for numerical experimentation and exploration as well as algebraic considerations, reflections, and manipulations that have demonstrated the power of algebra in giving insights into phenomena that are describable with numbers.

Reference

Kennedy, R. E., Goodman, T. A. and Best, C. H. (1981). 'Discovering patterns from a divisibility test', *School Science and Mathematics* 81, 34-36.

3.6 Dave Faulkner on 'Cycles'
Vol. 20, no. 5 (November 1991), pp 6-8

In this article I will look more closely at how to predict the length of cycles and discuss ideas which could form the basis for investigation.

Step I

Recap The cycles [in Gordon Haigh's article] were formed in the following way. First select multipliers p and q; suppose $p = 2$, $q = 5$. Then pick a number, say 39, and separate the rest (= X) from the units (= Y); this gives $X = 3$, $Y = 9$. Now compute $pX + qY$.

$$\begin{array}{ccc} & 3 & 9 \\ \times 2 \downarrow & & \downarrow \times 5 \\ & 6 + 45 & = 51 \end{array} \qquad 39 \longrightarrow 51.$$

Continuing in this way $39 \longrightarrow 51 \longrightarrow 15 \longrightarrow 27 \longrightarrow 39 \longrightarrow \cdots$ forming a cycle. Note that not all numbers are part of a cycle; for example 3 and 87 map to 15 and 51 respectively.

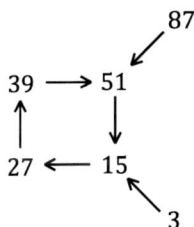

Further investigation

I worked out cycles for various combinations of p and q and found that some produced several short cycles while others included much longer ones.

Example 1

$p = 5$, $q = 2$ gives 3 cycles of length 4, one cycle of length 2 and one number which maps to itself.

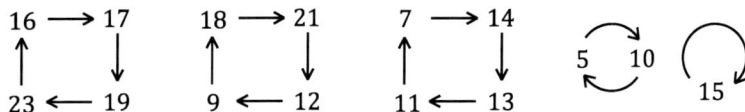

Example 2

$p = 3$, $q = 4$ leads to 2 cycles of length 18 and one number which maps to itself.

$$12 \longrightarrow 11 \longrightarrow 7 \longrightarrow 28 \longrightarrow 38 \longrightarrow 41 \longrightarrow 16 \longrightarrow 27$$

$$3 \qquad\qquad\qquad\qquad\qquad\qquad\qquad\qquad 34$$

$$10 \longleftarrow 21 \longleftarrow 33 \longleftarrow 36 \longleftarrow 9 \longleftarrow 30 \longleftarrow 26 \longleftarrow 25$$

$$22 \longrightarrow 14 \longrightarrow 19 \longrightarrow 39 \longrightarrow 45 \longrightarrow 32 \longrightarrow 17 \longrightarrow 31$$

$$24 \qquad\qquad\qquad\qquad\qquad\qquad\qquad\qquad 13$$

$$37 \circlearrowright$$

$$6 \longleftarrow 20 \longleftarrow 42 \longleftarrow 29 \longleftarrow 35 \longleftarrow 18 \longleftarrow 23 \longleftarrow 15$$

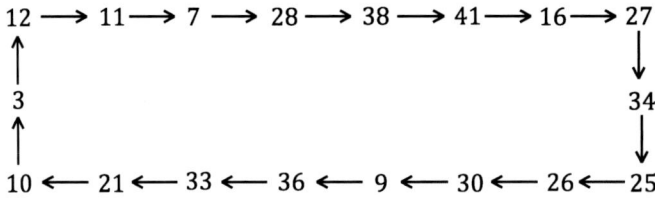

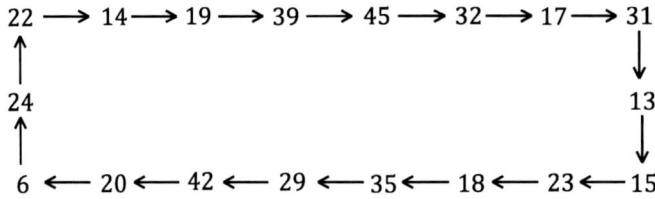

Background to method

I will take the case $p = 2, q = 5$ to outline the ideas that motivate my method for predicting cycle length. Suppose that $t_1, t_2, t_3, \ldots t_n$ are different numbers such that

$$t_1 \rightarrow t_2 \rightarrow t_3 \rightarrow \cdots t_n \rightarrow t_1 \tag{1}$$

Choose any number and express this as $10X + Y$ using the notation described earlier. This is mapped to $pX + qY$.

i. e. $\qquad\qquad 10X + Y \rightarrow pX + qY \tag{2}$

Note the identity

$$q(10X + Y) \equiv (10q - p)X + (pX + qY) \tag{3}$$

In the particular case under consideration $(p = 2, q = 5)$ these become

$$10X + Y \rightarrow 2X + 5Y \tag{2a}$$

and

$$5(10X + Y) \equiv 48X + (2X + 5Y) \tag{3a}$$

From (1), $t_r \rightarrow t_{r+1}$ (define $t_{n+1} = t_1$). Suppose $t_r = 10X + Y$ and then from $(2a)$, $t_{r+1} = 2X + 5Y$. (Note that X and Y vary for each r.)

Now substitute into $(3a)$, giving $5t_r = 48X + t_{r+1}$; therefore,

$$5t_r = t_{r+1} \pmod{48} \tag{4}$$

Putting $r = 1,2,3, \ldots n$ in (4) gives

$$5t_1 = t_2 \pmod{48}$$
$$5t_2 = t_3 \pmod{48}$$
$$\vdots \quad \vdots$$
$$5t_{n-1} = t_n \pmod{48}$$
$$5t_n = t_1 \pmod{48}$$

$$(5)$$

Taking the n equations in (5) and multiplying by $5^{n-1}, 5^{n-2}, \ldots 5^1, 5^0$ respectively yields

$$5^n t_1 = 5^{n-1} t_2 \pmod{48}$$
$$5^{n-1} t_2 = 5^{n-2} t_3 \pmod{48}$$
$$\vdots \quad \vdots$$
$$5^2 t_{n-1} = 5 t_n \pmod{48}$$
$$5 t_n = t_1 \pmod{48}$$

$$(6)$$

Adding the n equations in (6) and simplifying gives

$$5^n t_1 = t_1 \pmod{48} \tag{7}$$

And if $(t_1, 48) = f$, then

$$5^n = 1 \left(\mod \left(\frac{48}{f} \right) \right). \tag{8}$$

Application of method

Now we can use (8) to predict the length of the cycle to which a given number belongs or eventually enters. In the table on the next page we list the factors, f, of 48; list the N such that $(N, 48) = f$; substitute for f in (8); and find the smallest positive n for which (8) is true. This value of n predicts the cycle length.

This method has proved reliable in predicting the length of cycles for several other combinations of p and q. More work is needed to develop a general theory of cycles but I hope this will have pointed the way.

Step II

In the first step we were concerned with finding out how many cycles existed and what length they were. Now we will focus our attention on a particular cycle and the numbers which lead into it. The resulting diagram will be called a subchart.

111

f	N such that $(N, 48) = f$	$5^n = 1\left(\mathrm{mod}\left(\frac{48}{f}\right)\right)$	n	Cycles of length n
1	1, 5, 7, 11, 17, 19, 23, 25, 29, 31, 35, 37, 41, 43, 47, …	$5^n = 1 \ (\mathrm{mod}\ 48)$	4	$7 \to 35 \to 31 \to 11 \to 7$; $13 \to 17 \to 37 \to 41 \to 13$; $19 \to 47 \to 43 \to 23 \to 19$; $53 \to 25 \to 29 \to 49 \to 53$
2	2, 10, 14, 22, 26, 34, 38, 46, …	$5^n = 1 \ (\mathrm{mod}\ 24)$	2	$2 \leftrightarrow 10$; $14 \leftrightarrow 22$; $26 \leftrightarrow 34$; $38 \leftrightarrow 46$
3	3, 9, 15, 21, 27, 33, 39, 45, …	$5^n = 1 \ (\mathrm{mod}\ 16)$	4	$51 \to 15 \to 27 \to 39 \to 51$; $33 \to 21 \to 9 \to 45 \to 33$
4	4, 20, 28, 44, …	$5^n = 1 \ (\mathrm{mod}\ 12)$	2	$4 \leftrightarrow 20$; $28 \leftrightarrow 44$
6	6, 18, 30, 42, …	$5^n = 1 \ (\mathrm{mod}\ 8)$	2	$6 \leftrightarrow 30$; $18 \leftrightarrow 42$
8	8, 40, …	$5^n = 1 \ (\mathrm{mod}\ 6)$	2	$8 \leftrightarrow 40$
12	12, 36, …	$5^n = 1 \ (\mathrm{mod}\ 4)$	1	12; 36
16	16, 32, …	$5^n = 1 \ (\mathrm{mod}\ 3)$	2	$16 \leftrightarrow 32$
24	24, …	$5^n = 1 \ (\mathrm{mod}\ 2)$	1	24
48	48, …	$5^n = 1 \ (\mathrm{mod}\ 1)$	1	48

Example 1

Continue with $p = 2$, $q = 5$ as used in the previous step. We will look more closely at the numbers N such that $(N, 48) = 3$. They are

3	9	15	21	27	33	39	45
51	57	63	69	75	81	87	93
99	105	111	117	123	129	135	141
147	153	159	165	171	177	183	189
195	201	207	213	219	225	231	etc

These divide into two subcharts and we will consider the cycle $51 \rightarrow 15 \rightarrow 27 \rightarrow 39 \rightarrow 51$ and those numbers which map into it.

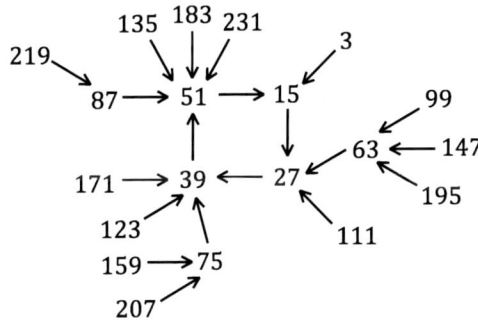

Our subchart only includes numbers up to 231 but it is sufficient to raise some interesting questions.

Question 1

It appears that some numbers, such as 3, are not mapped onto, i.e. they have no 'ancestors'. Is this true? Since $10X + Y \rightarrow 2X + 5Y$ we need to consider the equation $2X + 5Y = n$. The constraints on X and Y are such that 1 and 3 are the only numbers for which the equation is insoluble. So 1 and 3 are the only numbers with no ancestors.

Question 2

Which numbers map to 51? Have we found them all? Since $10X + Y \rightarrow 2X + 5Y$ we require X and Y such that $2X + 5Y = 51$. Y must be odd giving $Y = 1,3,5,7,9$ and $X = 23,18,13,8,3$ respectively. Therefore 231, 183, 135, 87 and 39 are the only five numbers which map to 51.

Question 3

What do you observe about the numbers which map to 51? They are equal mod 48. Such a relationship may seem surprising at first but it can be justified as follows. Suppose $t_1 \rightarrow 51$ and $t_2 \rightarrow 51$. Since $10X + Y \rightarrow 2X + 5Y$ and

$$5(10X + Y) \equiv 48X + (2X + 5Y)$$

then $5t_1 = 48X_1 + 51,$

and $5t_2 = 48X_2 + 51.$

Subtracting:

$5(t_1 - t_2) = 0 \pmod{48}$
$\Rightarrow \quad t_1 - t_2 = 0 \pmod{48}$
$\Rightarrow \qquad t_1 = t_2 \pmod{48}$.

Therefore, all numbers mapping onto 51 are equal mod 48.

Question 4

Which numbers map to themselves? Since $10X + Y \rightarrow 2X + 5Y$ we have to solve

$10X + Y = 2X + 5Y$
$\quad 2X = Y$.

Y must be even, giving $Y = 0, 2, 4, 6, 8$ and $X = 0, 1, 2, 3, 4$ respectively. Therefore, 12, 24, 36 and 48 are the only identities. Note that this agrees with the identities listed in the first step.

Question 5

Which numbers map to larger numbers? This question can be answered in a similar way to the previous one and is left as an exercise for you (or your students).

Conclusion

I believe that cycles offer many opportunities for investigative work. I hope that you will select other combinations of p and q and address the above questions together with those of your own.

Step III

My final step looks at multiplication tables.

Recap

$10X + Y \rightarrow pX + qY$.

If $p = 4, q = 1$, we have for example:

$$
\begin{array}{cccc}
 & 3 & 6 & \\
\times 4 \;\downarrow & & \downarrow\; \times 1 & \\
 & 12 & + \; 6 & = 18
\end{array} \qquad 36 \rightarrow 18.
$$

$$
\begin{array}{cccc}
 & 10 & 2 & \\
\times 4 \;\downarrow & & \downarrow\; \times 1 & \\
 & 40 & + \; 2 & = 42
\end{array} \qquad 102 \rightarrow 42.
$$

Example 1

Take $p = 4, q = 1$ and construct the subchart containing multiples of 6.

78
84
36 ← 90
96
18 ← 42 ← 102
48
etc.
6 ← 12 ← 24 ← 54
60
30 ← 66
72

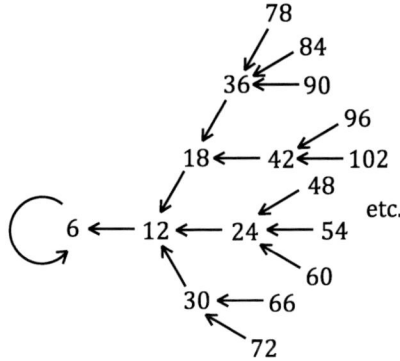

It appears that by listing the multiples of 6 and alternating the access routes between 3 and 2 branches we can construct part of the chart for $p = 4, q = 1$.

Example 2

The subchart for $p = 6, q = 1$ which contains multiples of 4 is shown below.

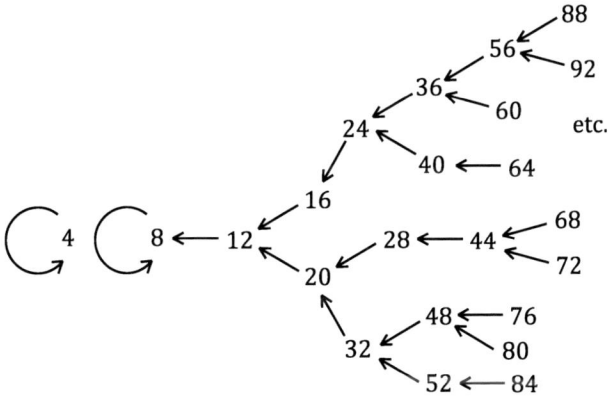

88
56 ←
92
36 ←
60
24 ←
etc.
40 ← 64
16
68
4   8 ← 12 ←   28 ← 44 ←
72
20
48 ← 76
32 ←
80
52 ← 84

The multiples of 4 emerge, in order, column by column. Can you describe a rule for continuing the subchart?

We will now consider the number of entries in each column. Ignoring the isolated 4 we get the sequence 1, 1, 2, 3, 5, 8 ... This might lead you to expect a Fibonacci sequence but...

Example 3

Take $p = 7, q = 1$ and produce the subchart containing multiples of 3. It begins with 3 cycles each containing just one number

3   6   9

and then 12, 15, 18, ... all eventually lead to the unit cycle containing 9. Can you identify a rule like that given in Example 1 which would enable you to continue the subchart?

If this does not present a sufficient challenge then carry out further investigations until you find one that does!

3.7 Tony Orton on 'The fascination of mathematics: Mappings'
Vol. 41, no. 5 (November 2012), pp 35-44; correction in Vol. 42, no. 1 (January 2013), p 2

Did you know that the mapping $x \rightarrow x^2 + 1$ (mod 100) leads to the pattern of Figure 1? I remember first seeing this diagram some 30 years ago, yet mathematics teachers may not be aware of this surprising result. Here is surely a fascinating sideline of secondary school mathematics, involving only relatively simple number work. But is there a broader and deeper investigation here?

Researching $x \rightarrow ax^2 + b$ (mod 100)

One obvious question for further research is whether any other mappings of the form $x \rightarrow ax^2 + b$ (mod 100) produce a connected pattern like Figure 1 and, if they do, whether they look exactly the same.

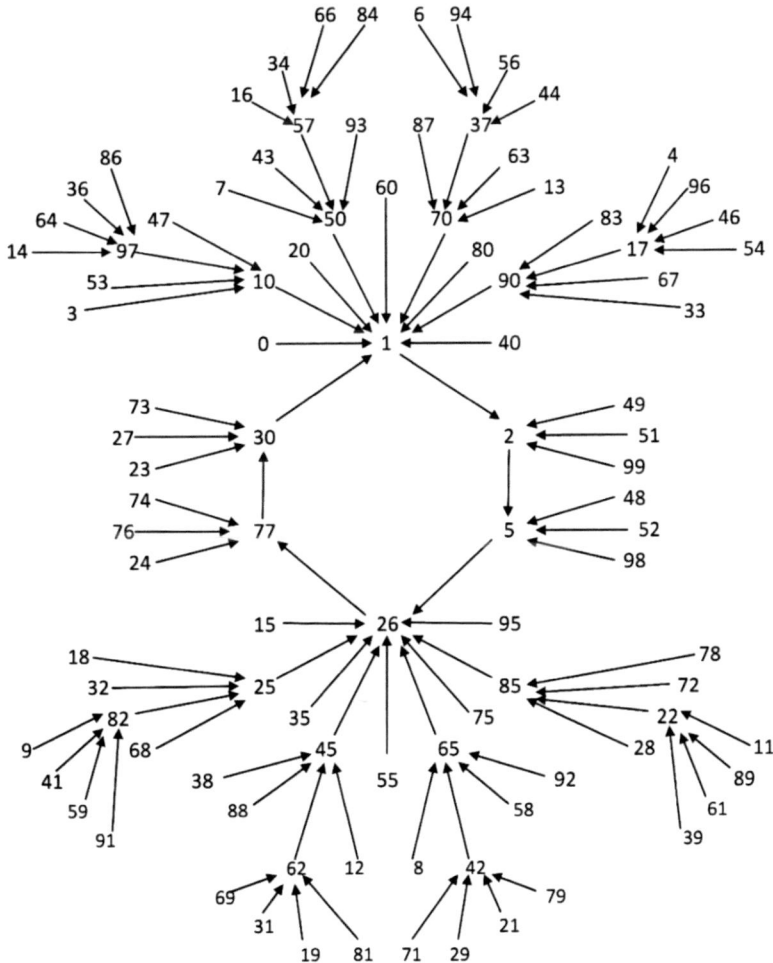

Fig. 1: $x \rightarrow x^2 + 1$ (mod 100)

Here, 'connected' is taken to mean that all the possible values of x (0 to 99 in mod 100) and their images in the mapping are linked into a single diagram. For simplicity at this stage, if we restrict ourselves to $1 < a < 9$ and $0 < b < 9$, our first example of miniaturizing the problem, we discover that $x \rightarrow 5x^2 + 1$ and $x \rightarrow 6x^2 + 1$ (Fig. 2), $x \rightarrow 3x^2 + 7$ and $x \rightarrow 4x^2 + 9$ are all examples of mappings which produce a connected pattern. Figure 2 reveals immediately that not all of the connected mappings look like Figure 1. One of these four further examples of connectedness is geometrically identical to the pattern for $x \rightarrow x^2 + 1$ (which one?), but the others are all different. However, many of the possible patterns generated from mappings of the form $x \rightarrow ax^2 + b$ (mod 100) are not connected. So, what do they all look like? What is special about the mappings which produce connected patterns? How many other distinct patterns emerge from using the whole range of possible values of a and b? What mappings, though algebraically different, nevertheless produce identical patterns? (*Figure 2 and all subsequent figures are shown at the end of the article.*)

At this point, it is necessary to introduce some terminology in order to shorten the explanations. The word 'pattern' has been adopted for the complete diagrammatic representation of a mapping, however many separate parts it has, and the word 'configuration' has been adopted for any of the separate parts within any overall non-connected pattern. The pattern for $x \rightarrow x^2 + 2$ (Fig. 3), for example, consists of four separate configurations, $x \rightarrow 4x^2 + 1$ (Fig. 4) has two configurations, and there are many other patterns which contain more than two configurations (Figs. 5 and 6). It should be clear by now that many of these patterns can form excellent display material. Note that Figures 2 and 3 use 'block format', which is often an easier way to record a pattern.

Simplifying the task

It then quickly made sense to tackle a more limited task at this stage, that is, to focus on mod 10, in other words to miniaturize the task in yet another way. The resulting 90 diagrams are quick and easy to draw. The first big surprise was that these 90 mappings produce only 16 distinct patterns. The full classification of these is in Table 1, and diagrammatic examples of all 16 patterns are in Figure 7. Of these 16 patterns, four are connected, and these four incorporate 25 of the 90 mappings. The number patterns within the algebraic descriptions of the mappings provide plenty of food for thought, and may possibly also provide a beginning towards understanding why some patterns are connected and others are not. Note that some of the patterns incorporate numbers which map to themselves ('sink' numbers, shown encircled); others incorporate pairs of numbers which map to each other (two-way associations, shown by a double-headed arrow)

The second big surprise occurred when at one point, and quite accidentally, I switched to an alternative mapping, $x \rightarrow (ax + b)^2$ instead of $x \rightarrow ax^2 + b$. We find that $x \rightarrow (ax + b)^2$ (mod 10) produces the same pattern as $x \rightarrow ax^2 + b$ (mod 10) for the same values of a and b, though the ten numbers are usually in different positions! Some examples are shown in Figure 8. However, other variations such as $x \rightarrow ax^2 + b^2$ (mod 10) do not generally produce the same pattern. Herein lies a parallel investigation.

From mod 10 to mod 100

Using the mappings in Table 1, the first remarkable finding from researching the mod 100 versions is that the same 16 groups revealed in the mod 10 research still hold. In other words, there are no changes to the locations of any of the 90 mappings, and there are no extra groups. At this stage it becomes very clear that without using mod 10 to help identify groups of mappings, it could have been incredibly difficult to analyse the mod 100 mappings. However, all is not straightforward, because there are some complications within the 16 mod 100 groups. Although half of the groups still have the same number of configurations, and only one connected mod 10 group has split into two configurations ($x \rightarrow 2x^2 + 1$), some mod 100 groups have more configurations than any in mod 10 (Table 2), with as many as 10 configurations for Group 14 (Fig. 9). Also, despite the conviction that there were still only 16 groups, peculiarities had emerged in some of the patterns at this stage which weren't understood until much later.

Completing the analysis

To complete the analysis it was necessary first to extend the range of values of b to $0 < b < 99$, and then of a to $1 < a < 99$. The quick and easy way to do this turned out to be to build on existing results in steps of 10. Thus, $4x^2 + 7$ led first to $4x^2 + 17$, $4x^2 + 27$, $4x^2 + 37$, ... and subsequently to $14x^2 + 7$, $24x^2 + 7$, $34x^2 + 7$, ... and so on until the original mapping becomes 100 mappings, constituting what might be referred to as the $4x^2 + 7$ subgroup. Not only do all of these 100 mappings lie within the same subgroup, but every group becomes extended 100-fold, and 9000 mappings have been classified. Critically, also, the structure of the 16 groups remains unchanged, and the situation concerning the peculiarities becomes completely clarified, as will be explained later.

Last of all, extending the values of a to $1 < a < 99$ introduced a new category of mapping, namely when $a = 10, 20, 30, ..., 90$. This leads to an additional 900 mappings, which are easy to draw, but which unfortunately result in a slightly messy classification problem. The result is five versions of a new and somewhat different type of pattern, all of which are connected. An example is shown in Figure 10. Thus, at this late stage in the analysis, it is now clear that the classification of these mappings requires a seventeenth group, and that there are once again four groups of connected patterns.

Alternatives and variations

With 9900 diagrams potentially available as data, the aforementioned peculiarities can now finally be seen to be of two types, described here either as alternatives or as variations on a basic pattern. Alternatives only arise when a is even, and depend on whether a is divisible by 4 or not. This is demonstrated by comparing Figures 2 and 11. Thus, the two alternatives in a group occur equally frequently. An important question here is whether a group with two alternative forms is still one single group, or whether it is two groups. This might be a debatable point, except that it can be argued strongly that both alternatives always possess the same major properties which one would use to describe the group and distinguish it from the other groups. Groups 2, 3, 9, 10, and 11 all have two alternative forms.

Variations appear to arise in one fifth of the mappings in certain groups and are most easily explained through examples. Firstly, consider the $4x^2 + 1$ subgroup of the Group 9 mappings. When $a = 4$ and $b = 11$, the configuration which is arranged around a pentagonal circuit takes what might be called a degenerate form, in which it splits into five separate matching parts with sinks (compare Figures 4 and 12). The values of a and b which produce this degenerate form in the $4x^2 + 1$ subgroup are listed in Table 3. Secondly, considering the $x^2 + 9$ subgroup of Group 13 (Fig. 5), the decagonal circuit configuration can degenerate into five separate matching parts with two-way associations, as illustrated by $x \rightarrow x^2 + 19$ (Fig. 13). Consult Table 3 for the values of a and b which produce this degeneration, seen in Groups 9, 13 and 15.

A completely different variation occurs when the blocks of 10s and 5s migrate from one of the configurations to another for some mappings (compare Figures 14 and 3). Table 3 again records the values of a and b which produce this variation in the $x \rightarrow x^2 + 2$ subgroup. Groups 3, 4, 6, 7, 8, 10, and 14 incorporate this kind of variation. In Groups 4 and 8 other blocks of numbers also migrate in the variation, and $a = 25$ and 75 both produce a special degenerate form. Groups 3, 9 and 10 also have two alternative forms, so four different diagrams are required to illustrate all the possible patterns.

Group 17 has been omitted from this consideration of alternatives and variations because its five patterns, which all differ slightly from each other, seem to defy such classification. However, there are similarities between the Group 17 patterns (most noticeable when $a = 50$) and the Groups 4 and 8 patterns (most noticeable when $a = 25$ or 75). Table 4 shows that it would require 39 diagrams to demonstrate all possible patterns with all their possible alternatives and variations within the 17 groups.

What started as an open investigation led to an attempt to classify and understand the structure of a coherent group of mappings, namely $x \rightarrow ax^2 + b$ (mod 100). But that is all. There is surely much more yet to be understood from the data available. And then, as with any good investigation, there is also surely much more beyond, for example not only the obvious $x \rightarrow (ax + b)^2$ (mod 100), but also $x \rightarrow ax^3 + b$ (mod 100), $x \rightarrow ax^2 + b$ (mod p), where p is any value you care to take, etc. Finally, a word of warning; such investigations can become addictive!

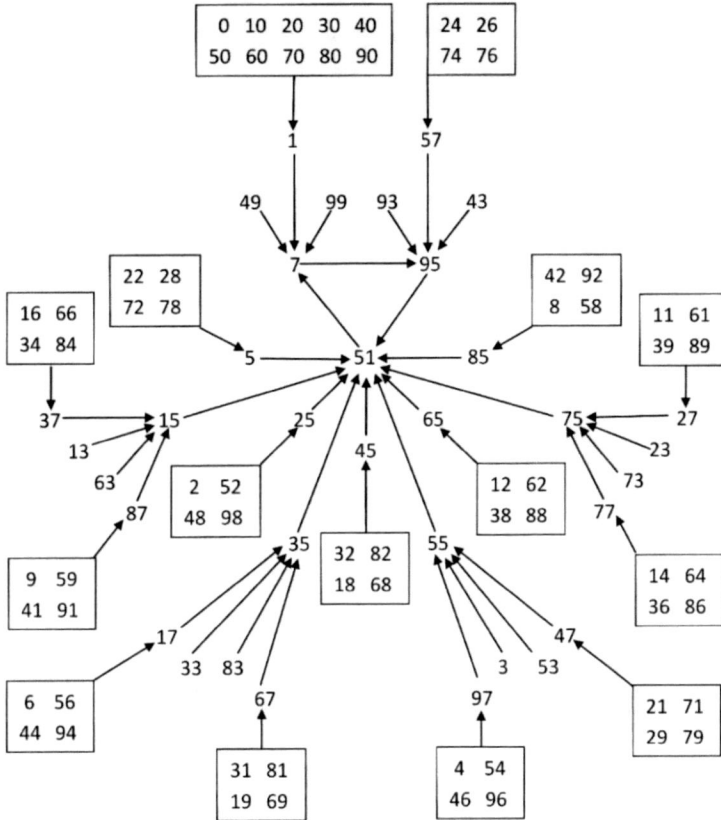

Fig. 2: $x \rightarrow 6x^2 + 1 \pmod{100}$

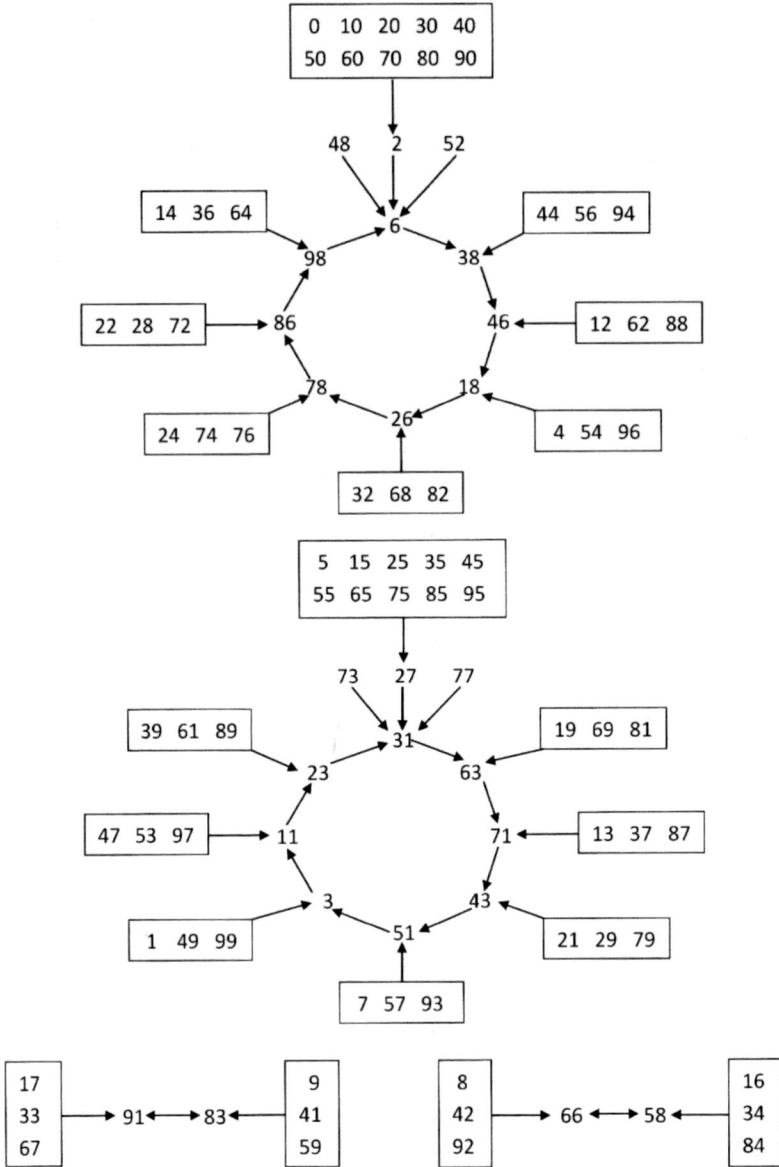

Fig. 3: $x \to x^2 + 2 \pmod{100}$

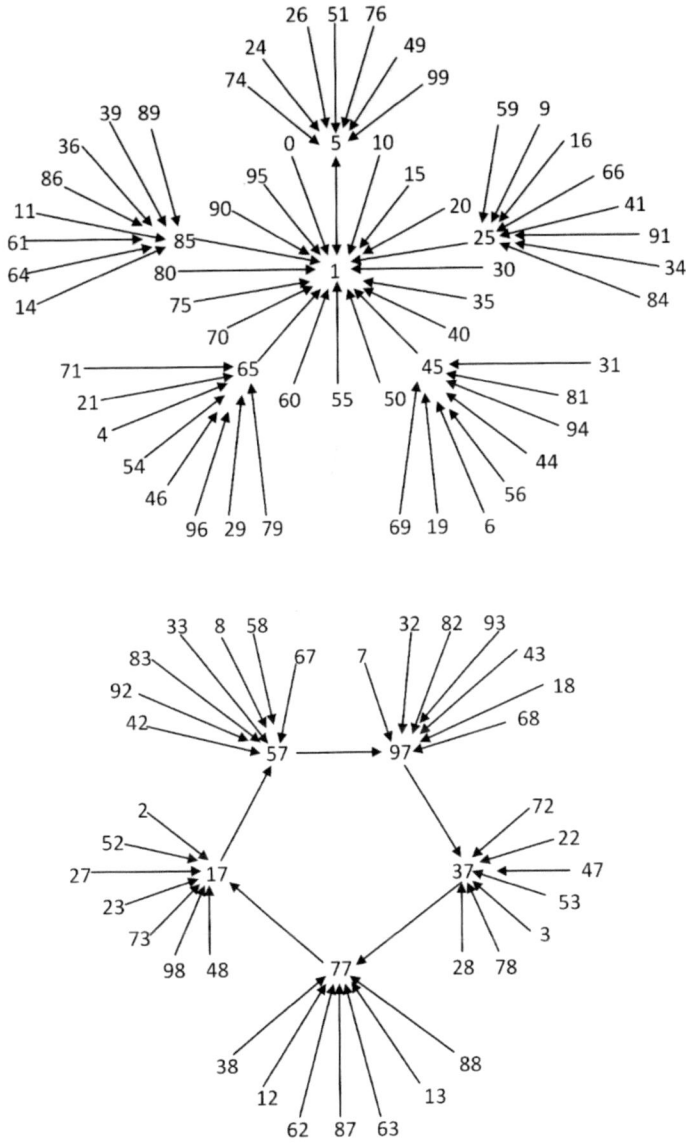

Fig. 4: $x \rightarrow 4x^2 + 2 \pmod{100}$

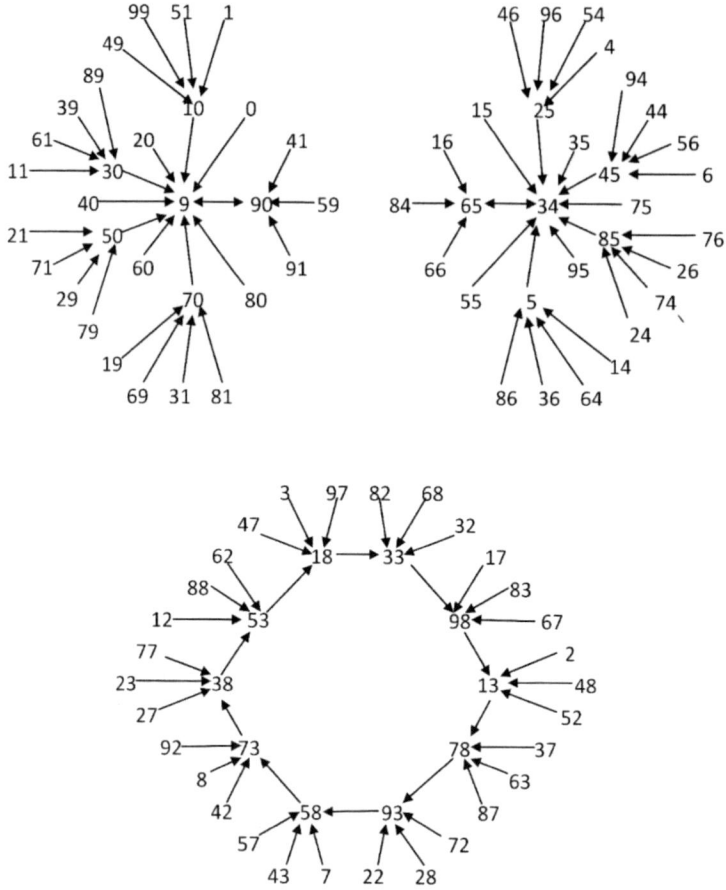

Fig. 5: $x \rightarrow x^2 + 9 \pmod{100}$

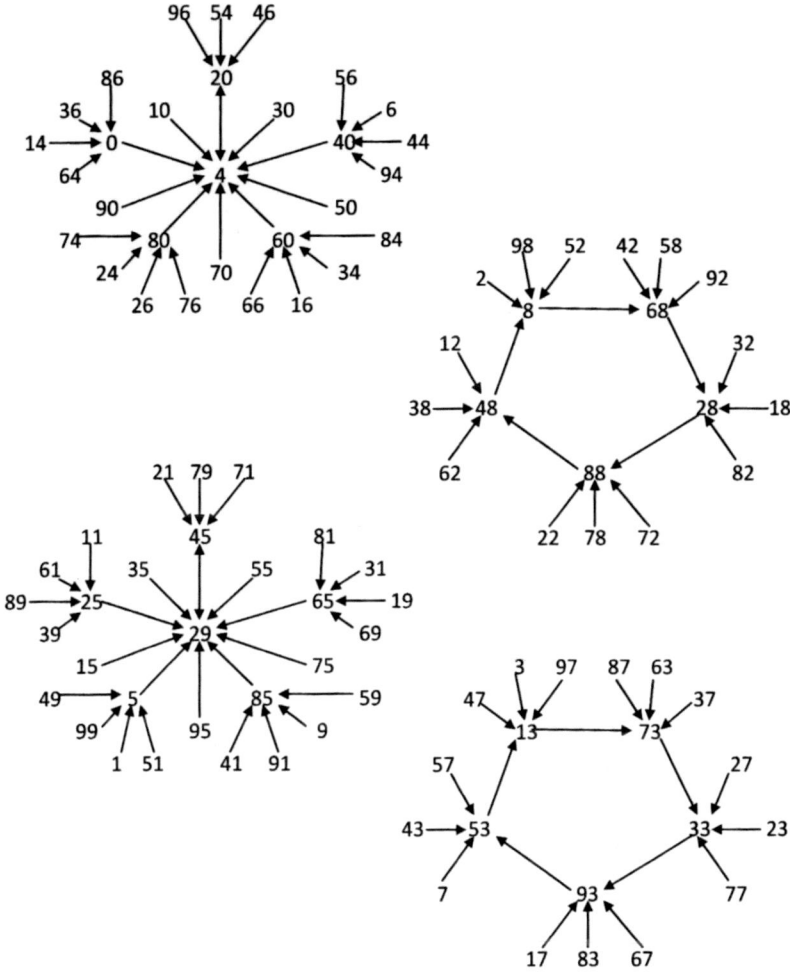

Fig. 6: $x \to x^2 + 4 \pmod{100}$

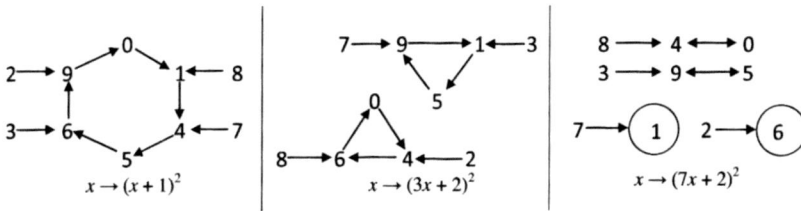

Fig. 8: Selected $x \to (ax + b)^2 \pmod{10}$ mappings for comparison with Fig. 7

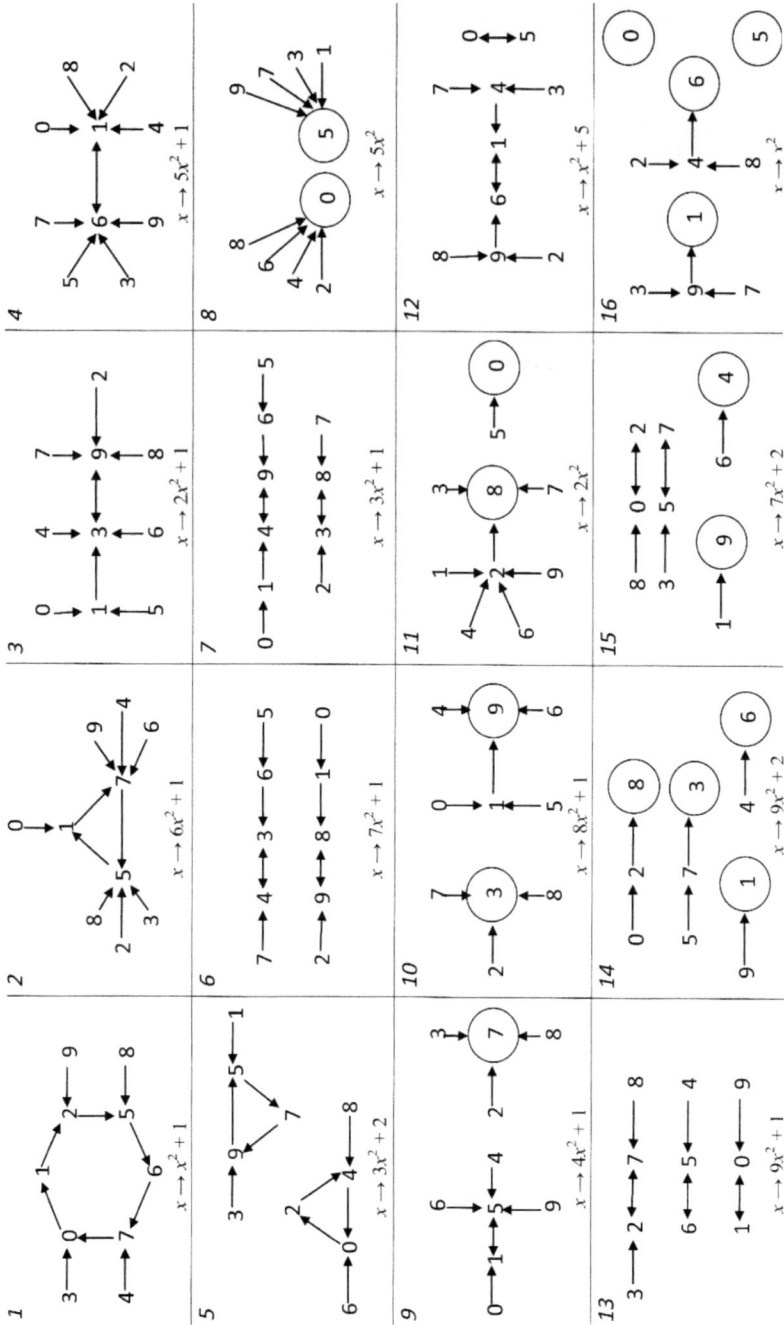

Fig. 7: The sixteen mod 10 patterns
(numbers within circles map to themselves)

126

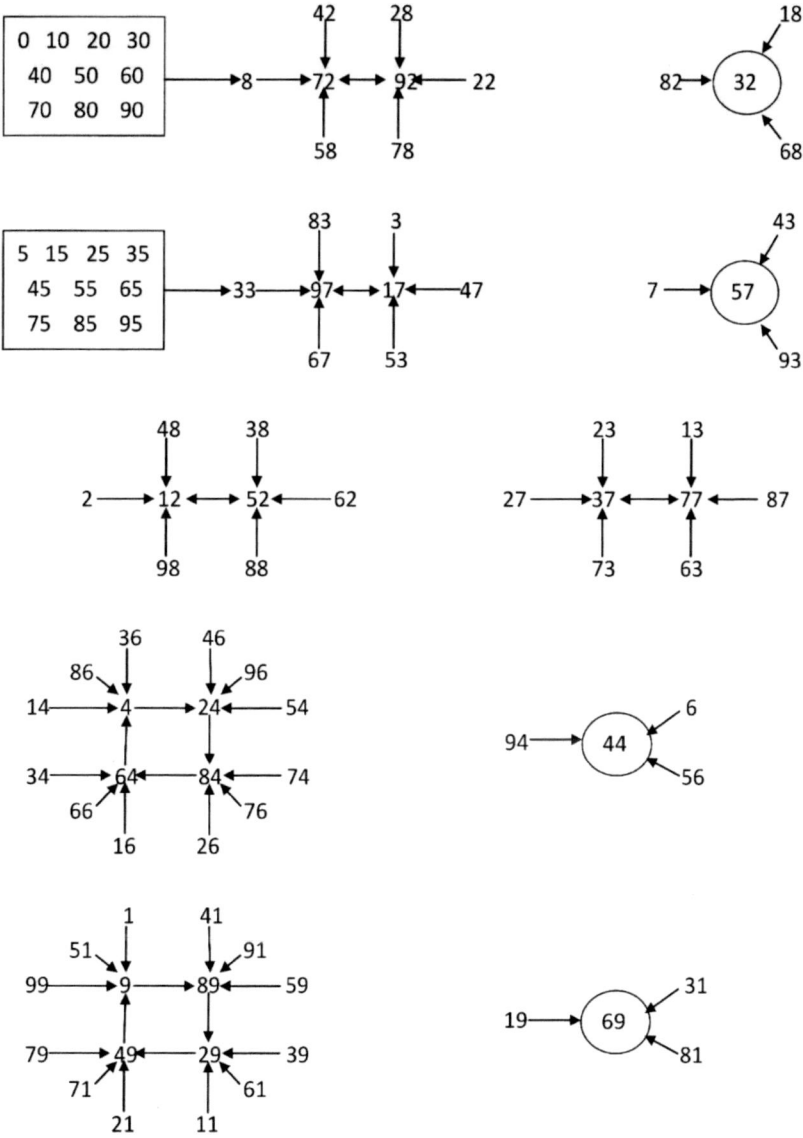

Fig. 9: $x \to x^2 + 8 \pmod{100}$

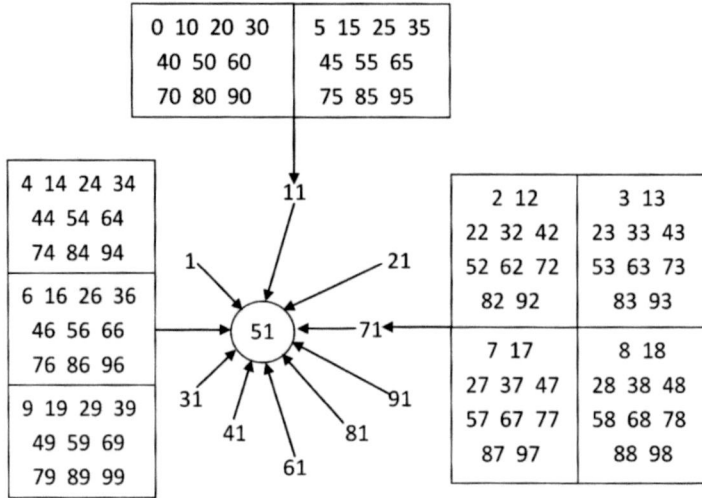

Fig. 10: $x \rightarrow 40x^2 + 11 \pmod{100}$

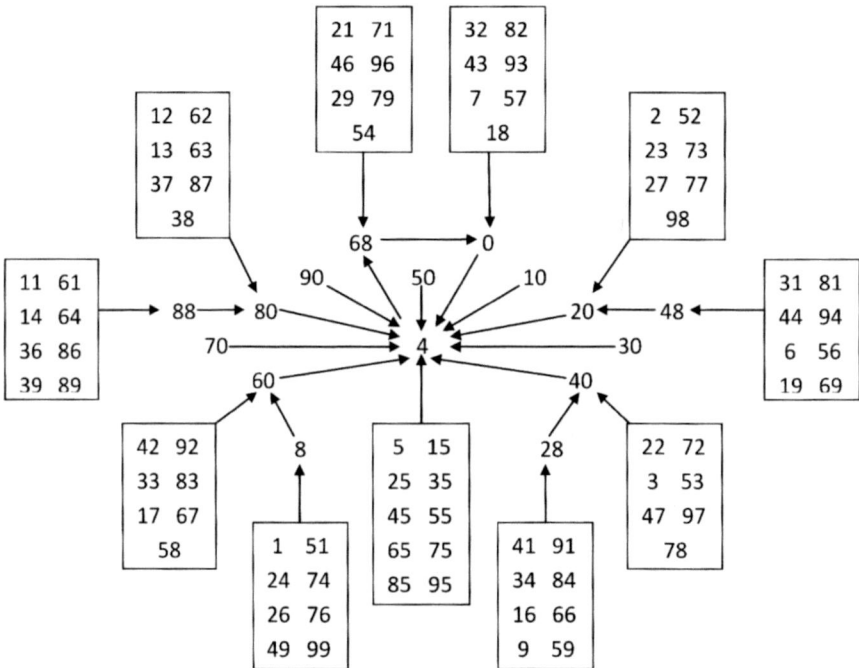

Fig. 11: $x \rightarrow 4x^2 + 4 \pmod{100}$

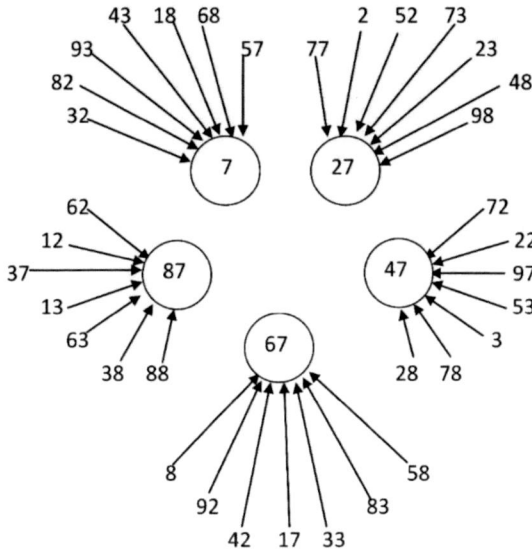

Fig. 12: The pentagonal configuration of $x \to 4x^2 + 114 \pmod{100}$

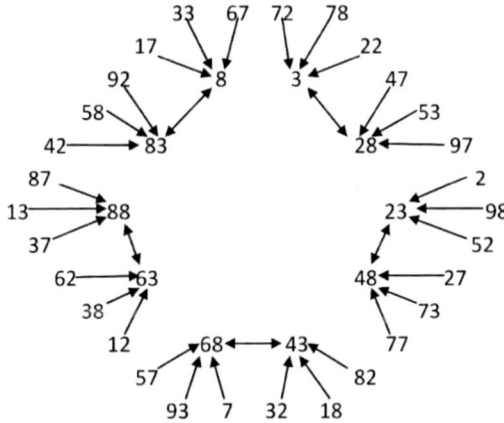

Fig. 11: The decagonal configuration of $x \to x^2 + 19 \pmod{100}$

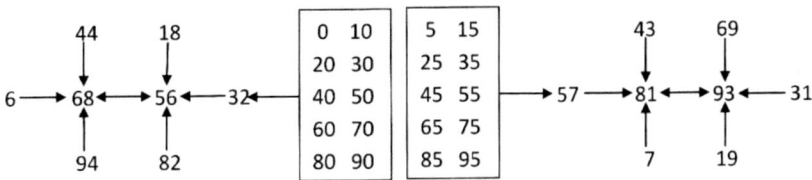

Fig. 14: The two string configurations of $x \to x^2 + 32 \pmod{100}$,
Showing the positions of the 5s and 10s block

Table 1: Classification of all $ax^2 + b$ (mod 10) mappings

Pattern group ref. no.	No. of configur- ations	$ax^2 + b$ (mod 10) mappings ($a \neq 0$) classified by geometrical representation	No. of mappings
1	1	$x^2 + 1, 7x^2 + 3, 3x^2 + 7, 9x^2 + 9$	4
2	1	$6x^2 + 1, 8x^2 + 2, 2x^2 + 3, 4x^2 + 4,$ $6x^2 + 6, 8x^2 + 7, 2x^2 + 8, 4x^2 + 9$	8
3	1	$2x^2 + 1, 2x^2 + 2, 4x^2 + 3, 8x^2 + 4,$ $2x^2 + 6, 6x^2 + 7, 4x^2 + 8, 8x^2 + 9$	8
4	1	$5x^2 + 1, 5x^2 + 3, 5x^2 + 5, 5x^2 + 7,$ $5x^2 + 9$	5
5	2	$3x^2 + 2, 9x^2 + 4, x^2 + 6, 7x^2 + 8$	4
6	2	$7x^2 + 1, x^2 + 2, 9x^2 + 3, 3x^2 + 4,$ $7x^2 + 6, x^2 + 7, 9x^2 + 8, 3x^2 + 9$	8
7	2	$3x^2 + 1, x^2 + 3, 9x^2 + 7, 7x^2 + 9$	4
8	2	$5x^2, 5x^2 + 2, 5x^2 + 4, 5x^2 + 6, 5x^2 + 8$	5
9	2	$4x^2 + 1, 2x^2 + 2, 8x^2 + 3, 6x^2 + 4,$ $4x^2 + 6, 2x^2 + 7, 8x^2 + 8, 6x^2 + 9$	8
10	2	$8x^2 + 1, 4x^2 + 2, 6x^2 + 3, 2x^2 + 4,$ $8x^2 + 6, 4x^2 + 7, 6x^2 + 8, 2x^2 + 9$	8
11	2	$2x^2, 4x^2, 6x^2, 8x^2,$ $2x^2 + 5, 4x^2 + 5, 6x^2 + 5, 8x^2 + 5$	8
12	2	$x^2 + 5, 3x^2 + 5, 7x^2 + 5, 9x^2 + 5$	4
13	3	$9x^2 + 1, 3x^2 + 3, 7x^2 + 7, x^2 + 9$	4
14	4	$9x^2 + 2, 7x^2 + 4, 3x^2 + 6, x^2 + 8$	4
15	4	$7x^2 + 2, x^2 + 4, 9x^2 + 6, 3x^2 + 8$	4
16	4	$x^2, 3x^2, 7x^2, 9x^2$	4
		TOTAL MAPPINGS	90

Table 3: Values of a and b which produce variations

$x \rightarrow 4x^2 + 1$ subgroup				$x \rightarrow x^2 + 9$ subgroup				$x \rightarrow x^2 + 2$ subgroup			
a	b	a	b	a	b	a	b	a	b	a	b
4	11	54	11	1	19	51	19	1	32	51	32
4	61	54	61	1	69	51	69	1	82	51	82
14	21	64	21	11	29	61	29	11	12	61	12
14	71	64	71	11	79	61	79	11	62	61	62
24	31	74	31	21	39	71	39	21	42	71	42
24	81	74	81	21	89	71	89	21	92	71	92
34	41	84	41	31	49	81	49	31	22	81	22
34	91	84	91	31	99	81	99	31	72	81	72
44	1	94	1	41	9	91	9	41	2	91	2
44	51	94	51	41	59	91	59	41	52	91	52

Table 2: mod 10 and mod 100 configurations

Table 4: Different forms of the 17 patterns

Group ref. no.	Number of configurations	
	mod 10	mod 100
1	1	1
2	1	1
3	1	2
4	1	1
5	2	2
6	2	4
7	2	8
8	2	2
9	2	2
10	2	5
11	2	3
12	2	4
13	3	3
14	4	10
15	4	4
16	4	6

Group number	No. of configur-ations	Number of alternatives & variations on basic pattern
1	1	1
2	1	2
3	2	4
4	1	2
5	2	1
6	4	2
7	8	2
8	2	2
9	2	4
10	5	4
11	3	2
12	4	1
13	3	2
14	10	2
15	4	2
16	6	1
17	1	5
	TOTAL	39

4   *Geometric Counting*

4.0 Introduction

Introducing the idea of an investigation to a class of young mathematics' students can often be a challenge. A good approach is to pose a problem and see what happens.

Imagine the following scenario that has probably been played out in many schools.

A class of year 7 students are sitting expectantly in their first mathematics lesson in their 'new' secondary school. The teacher stands at the front with an 8 × 8 'chessboard' (probably on the electronic whiteboard) and asks the students how many squares they can see? Up go the hands – a student gives the answer of 64 – smiles and nods all round as they triumphantly demonstrate their skill with the 8-times table.

The smiles turn to confusion when the teacher says "no, there are more than 64"!

A couple of students raise their hands and identify the whole chessboard as a square. Groans follow from other students as they realise the answer is 65!

"No, there are more than 65", the teacher responds with a sly smile!

After a period of silence, a student will offer the notion of 2 × 2 squares and the class is off and running with the challenge of finding the number of 2 × 2 squares, 3 × 3 squares and so on. The investigation has begun.

This chapter offers several similar investigations which fall under the heading of 'geometric counting' or 'enumerative geometry'. Our first five articles are about counting the number of simple two-dimensional shapes within a larger two-dimensional shape: rectangles within a rectangle, triangles within a triangle etc. Going beyond just counting, as our theoretical group of year 7 students might have done, several articles explore algebraic proof of generalizations from simple shapes.

We then offer an article by Chris Pritchard on an investigation involving the number of cuboids within a cube. The penultimate article in this chapter demonstrates the opportunities of developing a challenging algebraic proof of the formula Chris finds. The final article draws together the two-dimensional world of 'a square's rectangles' and the three-dimensional world of 'a cube's cubes'.

4.1 Chris Pritchard on 'The number of oblongs on a chessboard'
Vol. 45, no. 1 (January 2016), pp 30-31

We know what squares and rectangles are, but sometimes the word 'oblong' is used loosely in place of 'rectangle'. All squares are rectangles because they possess the properties of the rectangle: each angle is right and both pairs of opposite sides are equal and parallel. Some rectangles are squares, but those which are not are the shapes which are correctly termed 'oblongs'.

On this understanding,

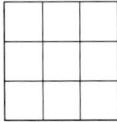

is a square,

is an oblong,

and both are rectangles.

The problem of finding the number of squares in a larger square is well known. Sometimes it is presented as the 'chessboard problem': on a chessboard, how many squares are there of all sizes? Some teachers simply ask how many squares there are, and when the answer of 64 comes back quickly, they say 'No', and the figure is rapidly raised to another incomplete answer, 65, as pupils include the square forming the perimeter. It is only at this time that they say 'of all sizes'.

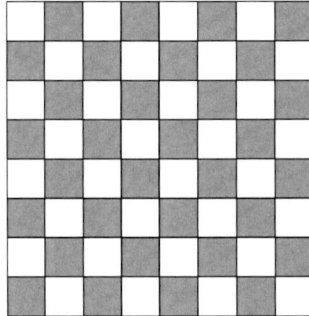

Clearly, the problem needs to be broken down into smaller steps. Perhaps the best place to start is to consider 2 × 2 squares. In the figure below, a square of black card has been placed on the board to show one possibility:

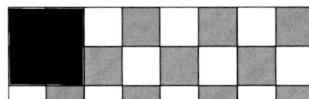

We could slide the card to the right to give:

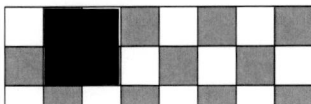

And, indeed there would be 7 positions overall as we move across the top of the board. There are also seven rows upon which we could place such cards. So there are $7 \times 7 = 49$ occurrences of squares of this size. If we cover the board with 3×3 squares, we find there are $6 \times 6 = 36$ of them, and so on. The solution is completed by putting this information into a table and summing:

Size	Occurrences	Totals
1×1	8×8	64
2×2	7×7	49
3×3	6×6	36
4×4	5×5	25
5×5	4×4	16
6×6	3×3	9
7×7	2×2	4
8×8	1×1	1
		204

There are 204 squares of all sizes on a standard chessboard.

Now to generalise: how many squares are there of all sizes on a $n \times n$ chessboard? We simply need to sum the squares of the positive integers to n. Labelling this sum T_s, we quote the standard result :

$$T_s = \sum_{r=1}^{n} r^2 = \frac{1}{6}n(n+1)(2n+1).$$

Now, let's consider how many rectangles there are on a square board, beginning with small boards. Here's a 2×2 board and its constituent rectangles:

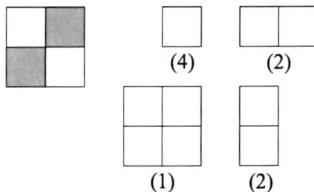

The bracketed numbers beneath each figure indicate the number of occurrences of such rectangles within it. The total number of rectangles is 9.

The equivalent analysis for the 3×3 board is shown alongside. Now the total number of rectangles is 36.

(9)   (6)   (6)

(4)   (3)   (3)

(2)

(2)

(1)

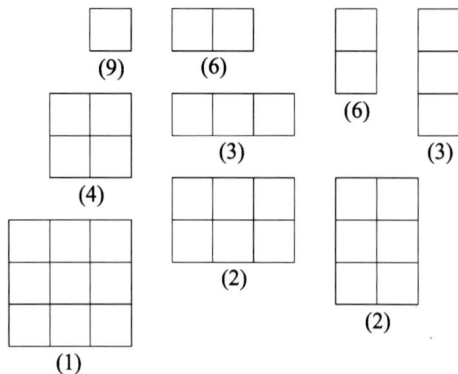

And again for the 4 × 4 board:

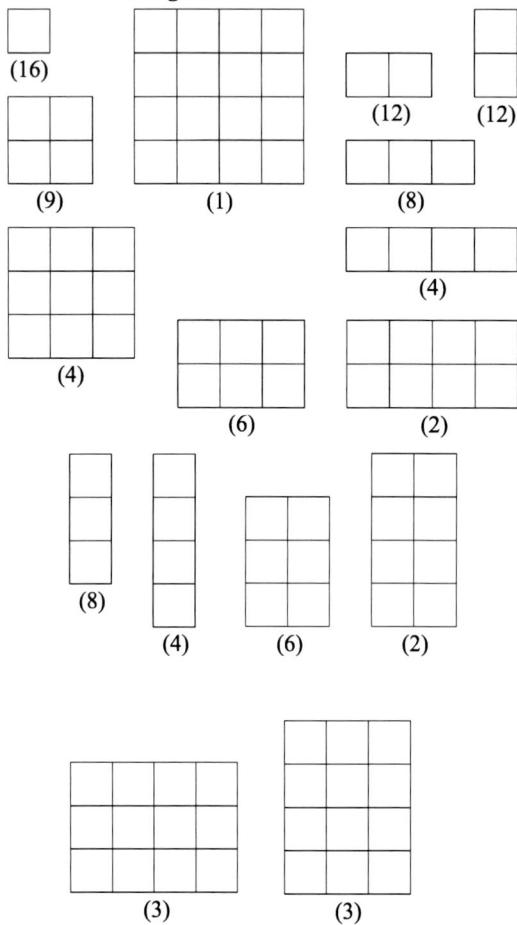

(16)   (12)   (12)

(9)   (1)   (8)

(4)

(4)

(6)   (2)

(8)

(4)   (6)   (2)

(3)   (3)

Here, the total number of rectangles is 100.

Perhaps it is no coincidence that all three totals generated to this point are perfect squares! Arranging the results for the 4 × 4 board in a table could help here:

		1	2	3	4
	1	16	12	8	4
Number	2	12	9	6	3
of rows	3	8	6	4	2
	4	4	3	2	1

Number of columns

The number of occurrences of true squares are shown on the leading diagonal, the number of occurrences of 2 × 1 rectangles (i.e. 12) is shown in the shaded cell, and so on. Working up from the bottom, row by row, the sixteen entries can be written as:

$$(1 + 2 + 3 + 4) + 2(1 + 2 + 3 + 4) + 3(1 + 2 + 3 + 4) + 4(1 + 2 + 3 + 4).$$

This factorises to give:

$$(1 + 2 + 3 + 4)(1 + 2 + 3 + 4) = (1 + 2 + 3 + 4)^2.$$

In a similar way, the total number of rectangles in a 3 × 3 square can be given as $(1 + 2 + 3)^2$ and in a 2 × 2 square as $(1 + 2)^2$.

It appears that the total number of rectangles in a square, T_r say, is always the square of a sum of positive integers. And remember that the total number of squares in a square, T_s, has already been shown to be the sum of squares of positive integers. To summarise, we have:

$$T_s = \sum_{r=1}^{n} r^2 = \frac{1}{6}n(n + 1)(2n + 1).$$

$$T_r = \left(\sum_{r=1}^{n} r\right)^2 = \left[\frac{1}{2}n(n + 1)\right]^2 = \frac{1}{4}n^2(n + 1)^2.$$

It is also worth noting that the latter result is the sum of the cubes.

To find the total number of oblongs in a square, T_b, we could go back to the diagrams. Alternatively, we could take away the number of squares from the number of rectangles:

$$T_b = \frac{1}{4}n^2(n + 1)^2 - \frac{1}{6}n(n + 1)(2n + 1)$$

which tidies up nicely to

$$\frac{1}{12}(n - 1)n(n + 1)(3n + 2).$$

Hence, the number of oblongs on a chessboard is:

$$\frac{1}{12} \times 7 \times 8 \times 9 \times 26 = 42 \times 26 = 1092.$$

4.2 'Placing four tiles 2 × 1 on a 4 × 4 square' by Beryl Boyd
Vol. 11, no. 1 (January 1982), pp 22-23

Part I: ways of placing four tiles 2 × 1 on a 4 × 4 square

How many different ways can you place four tiles 2 × 1 on a 4 × 4 square? Assumed rules: Tiles must not overlap, and each tile must occupy two whole squares vertically or horizontally, e.g. not

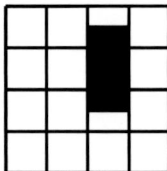

I started marking out 4 × 4 squares on squared paper and colouring possible positions of tiles. When I had made about 40 patterns, I stopped to review the results. Some were rotations or reflections of others; should I count these as different? Because I had coloured the positions of the tiles, the direction in which they had been placed was ambiguous in some cases, e.g.

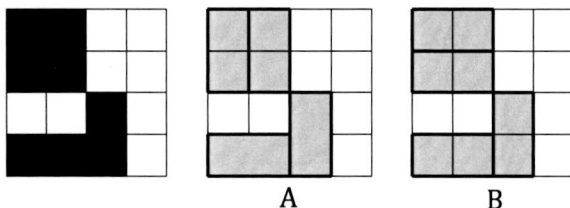

A   B

Had the square been made by A or B?

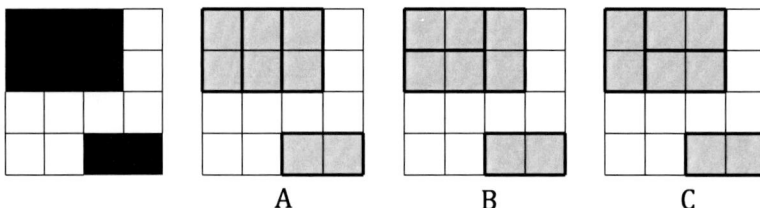

A   B   C

Had the rectangle been made by A or by B or by C?

In other words, each time I made a square from two tiles or a rectangle from three tiles, I needed to make it clear in which direction I had placed them and also to include any possible variations. From this stage onwards I decided to shade rather than colour. Groups of patterns were emerging, for example, those including

- a 2 × 2 square
- a 3 × 2 rectangle
- a 4 × 1 rectangle.

How many different patterns were there going to be? Would I be able to tell when I had found them all? I decided to make a chart, numbering the possible positions of the rectangular tiles.

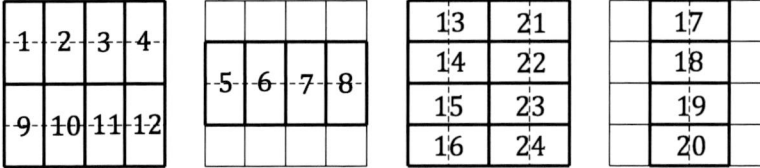

| 1 2 3 4 |
| 9 10 11 12 |

| 5 6 7 8 |

13	21
14	22
15	23
16	24

| 17 |
| 18 |
| 19 |
| 20 |

I started to list possible number combinations:

(1, 2, 3, 4)
(1, 2, 3, 5) – not possible because 5 would overlap 1
(1, 2, 3, 6) – not possible because 6 would overlap 2
(1, 2, 3, 7) – not possible because 7 would overlap 3
(1, 2, 3, 8)

I stopped, and made a list of impossible combinations

1 would overlap 5, 13, 14
2 would overlap 6, 13, 14, 17, 18
3 would overlap 7, 17, 18, 21, 22
4 would overlap 8, 21, 22
5 would overlap 9, 14, 15
6 would overlap 10, 14, 15, 18, 19
7 would overlap 11, 18, 19, 22, 23
8 would overlap 12, 23, 24
9 would overlap 15, 16 and so on.

I resumed work on my list of possible number combinations, omitting the impossible ones or crossing them out if I had included them in error.

(1, 2, 3, 4), (1, 2, 3, 8), (1, 2, 3, 9), (1, 2, 3, 10), (1, 2, 3, 11), (1, 2, 3, 12)...

Several days and two exercise books later, I reached (21, 22, 23, 24).

I started to draw the patterns beginning with the most obvious outline, a rectangle 4 × 2.

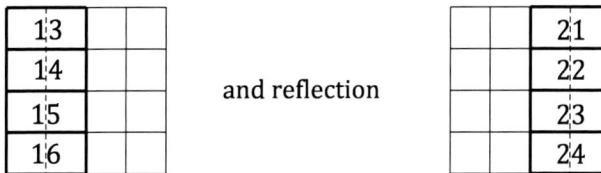

13	
14	
15	
16	

and reflection

	21
	22
	23
	24

(13, 14, 15, 16)   (21, 22, 23, 24)

by rotating them a quarter turn

(1, 2, 3, 4)   (9, 10, 11, 12)

I crossed out these sequences from my list, noting that by rotating the patterns a half-turn the sequences would recur. I investigated different ways of making the same rectangle and crossed off the combinations representing the identity, rotation and reflection of these.

Then I made the position of the rectangle central along one side; this led to ways of representing

two squares 2 × 2
two rectangles 4 × 1
one rectangle 3 × 2 and one rectangle 2 × 1
one square 2 × 2 and two rectangles 2 × 1
one rectangle 4 × 1 and two rectangles 2 × 1.

I drew all of them with their reflections, and crossed off the number combinations which they and their rotations represented. At this stage I was left with many more unrepresented number combinations than I had anticipated, and I had a file full of patterns.

Part II: ways of placing four rectangles 2 × 1 on a square 4 × 4, without making a square or larger rectangle

As I came to patterns which were the same if rotated, I collected them together and worked through the rest of my number combinations. (The supply of squared paper was a recurring frustration.)

At the end I had drawn more than 600 patterns (many were reflections) and with their rotations they represented more than 2500 number combinations. I kept a record of the numbers representing each pattern, in a book with pages numbered the same as the file of patterns. I found the 'perfect' pattern, with its vertical, horizontal and diagonal symmetry.

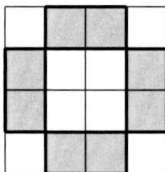

(5, 8, 17, 20) Rotation = reflection = identity.

Activities for Children (tried by the author in primary classes)

Draw a square 4 × 4 (e.g. on 2 cm squared paper). Cut out four rectangles 2 × 1 from a separate piece of paper. Colour your rectangles one colour.

1. Place the rectangles within the square so that each rectangle covers two whole squares, and the rectangles do not overlap. Record the position of your rectangles by shading part of the square. Repeat several times,

changing the position of the rectangles each time. What fraction of the 4 × 4 square have you shaded each time? What area have you shaded each time?

2. Work with a partner. One person places three of the rectangles according to previous rules. In how many different positions is it possible for the second person to place the fourth rectangle? How can the first person make it easier or more difficult for the second person to find several ways?

3. Work with a partner – two rectangles each. The first person places two rectangles within square. The second person must place his/her rectangles so that the total shape is symmetrical in some way.

Where to go next? What would happen using a different size square and/or different size rectangles and/or different number of rectangles. What would happen using rhombi within an equilateral triangle? What would happen if ... ?

4.3 David Pagni on 'Counting triangles with triangle numbers

Vol. 44, no. 5 (November 2015), pp 20-22

The following drawing is often presented in mathematics textbooks with the directions to find all triangles (Fig. 1).

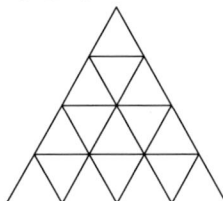

Fig. 1: Triangle of triangles

Students use various strategies to identify and count the triangles, including counting by triangle size. If we let the small triangles have sides of unit length, we count 16 small triangles, seven triangles with sides of length 2, three with sides of length 3, and one with side of length 4. When counting triangles with a base length of 2, one needs to not only 'see' the overlapping triangles, but the 'point up' and 'point down' triangles as well (Figs. 2 and 3).

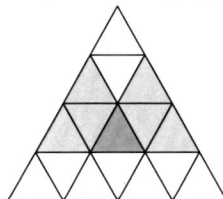

Fig. 2: Overlapping 'point up' triangles of side length 2

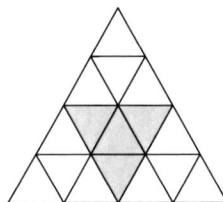

Fig. 3: 'Point down' triangle of side length 2

Equivalently, some students ascribe unit area to the small triangles and thus arrive at the same count: 16 triangles of area 1, seven triangles of area 4, three triangles of area 9, and one triangle of area 16 (Table 1). As an aside, note the pattern of perfect squares for triangle area. Students may see this and want to investigate. Hopefully they will notice that each time the side length is increased by one unit, a new row of small triangles is added to the area. Since the new row always contains an odd number of small triangles, each area measure is the sum of the odd numbers, beginning with 1. Thus, the triangle with area 16 is made of $1 + 3 + 5 + 7$ small triangles and we know the sum of the first four odd numbers is equal to 4^2.

Triangle side length	Triangle area	Number of triangles
1	1	16
2	4	7
3	9	3
4	16	1

Table 1: Counting triangles

Realizing that categorizing triangles by length of side or area are equivalent methods for counting triangles, we might notice that orientation can differ. Counting 'point up' and 'point down' triangles leads to the following results (Table 2).

Triangle base length	Point up	Point down
1	$1 + 2 + 3 + 4 = 10$	$1 + 2 + 3 = 6$
2	$1 + 2 + 3 = 6$	$1 = 1$
3	$1 + 2 = 3$	0
4	$1 = 1$	0
Total	20	7

Table 2: Triangle counts by orientation

Here is where the pattern of triangle numbers emerges! Recall that triangle numbers, T_n, are of the form

$$1 + 2 + 3 + \cdots + n = \frac{n(n+1)}{2}.$$

We investigate this relationship further and pursue a generalization for the numbers of triangles of different side lengths and total number of triangles. We do this by finding all triangles for the following shape (Fig. 4). Using what we learned from the previous example (Fig. 1 and Table 2), we count triangles by size and orientation, 'point up' or 'point down' (Table 3). The 'point up' numbers seem to follow a nice triangle number pattern; that is, the total number of triangles with side of length k is $T_{n-(k-1)}$. For example, there are

$$T_{8-(1-1)} = T_8 = \frac{8 \times 9}{2} = 36$$

'point up' triangles with side length 1. This is a pattern that is easily discernible by students researching this problem. The number of 'point down' triangles with side of length k is not so obvious, but can still be found with some algebra. The numbers appear to be odd triangle numbers, that is, triangle numbers of the form $T_{n-(2k-1)}$. For example, from Table 3, there are $T_{8-(2 \times 2-1)} = T_5 = 15$ 'point down' triangles with side length 2. To research this problem further we need to see what happens when we have an odd number of rows in the original triangle (Fig. 5).

Triangle side length	Point up	Point down
1	$T_8 = 36$	$1 + 2 + 3 + 4 + 5 + 6 + 7 = 28$
2	$T_7 = 28$	$1 + 2 + 3 + 4 + 5 = 15$
3	$T_6 = 21$	$1 + 2 + 3 = 6$
4	$T_5 = 15$	$1 = 1$
5	$T_4 = 10$	0
6	$T_3 = 6$	0
7	$T_2 = 3$	0
8	$T_1 = 1$	0
Total	120	50

Table 3: Results for a triangle with eight rows

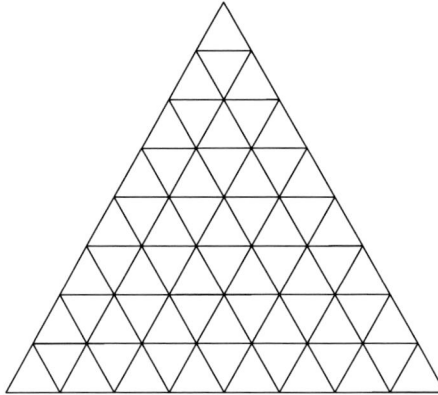

Fig. 4: Triangle with $n = 8$ rows of small triangles

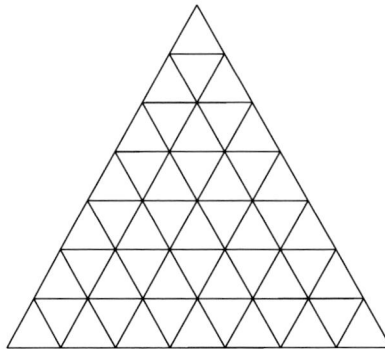

Fig. 5: Triangle with $n = 7$ rows of small triangles

Again, we record the results in a table (Table 4).

Triangle base length	Point up	Point down
1	$T_7 = 28$	$1 + 2 + 3 + 4 + 5 + 6 = 21$
2	$T_6 = 21$	$1 + 2 + 3 + 4 = 10$
3	$T_5 = 15$	$1 + 2 = 3$
4	$T_4 = 10$	0
5	$T_3 = 6$	0
6	$T_2 = 3$	0
7	$T_1 = 1$	0
Total	84	34

Table 4: Results for a triangle with seven rows

This time the numbers of 'point up' triangles follow the same pattern, but the numbers of 'point down' triangles are 'even' triangle numbers, i.e., triangle numbers of the form $T_{n-(2k-1)}$. For example, Table 4 shows that there are $T_{7-(2\times1-1)} = T_6 = 21$ 'point down' triangles with side length 1. Thus, for original triangle of even or odd number of rows, the number of 'point down' triangles with side length k is $T_{n-(2k-1)}$ and the number of 'point up' triangles is $T_{n-(k-1)}$.

So far the investigation can be carried out by students with only pattern recognition and can be generalized for individual length of triangle side with algebra. The final stage is more complex as we attempt to find a formula for any number of rows n. To help analyse our patterns we construct a table of numbers of triangles by length of side for each n (Table 5).

	Side length								Total triangles	
n	1	2	3	4	5	6	7	8	Up	Down
8	T_8	T_7	T_6	T_5	T_4	T_3	T_2	T_1	120	50
7	T_7	T_6	T_5	T_4	T_3	T_2	T_1		84	34
6	T_6	T_5	T_4	T_3	T_2	T_1			56	22
5	T_5	T_4	T_3	T_2	T_1				35	13
4	T_4	T_3	T_2	T_1					20	7
3	T_3	T_2	T_1						10	3
2	T_2	T_1							4	1
1	T_1								1	0

Table 5: Number of triangles by side length written as a triangle number

Step 1

For all 'point up' triangles we are seeking a formula for $T_1 + T_2 + \cdots + T_n$.

Since $T_i = \dfrac{i(i+1)}{2}$, we want

$$\sum_{i=1}^{n} \frac{i(i+1)}{2} = \frac{1}{2}\sum_{i=1}^{n}(i^2+i)$$

$$= \frac{1}{2}\left[\sum_{i=1}^{n}i^2 + \sum_{i=1}^{n}i\right]$$

$$= \frac{1}{2}\left[\frac{n(n+1)(2n+1)}{6} + \frac{n(n+1)}{2}\right]$$

$$= \frac{n(n+1)(2n+1+3)}{12}$$

$$= \frac{n(n+1)(n+2)}{12}.$$

Step 2

For all 'point down' triangles when n is even we want $T_1 + T_3 + T_5 + \cdots + T_{2i-1}$ for $i = 1,2,3,\dots,\dfrac{n}{2}$.

$$\sum_{i=1}^{n/2} T_{2i-1} = \frac{1}{2}\sum_{i=1}^{n/2}(2i-1)(2i) = \sum_{i=1}^{n/2}(2i^2-i)$$

$$= \frac{2\left(\frac{n}{2}\right)\left(\frac{n}{2}+1\right)\left(2\cdot\frac{n}{2}+1\right)}{6} - \frac{\left(\frac{n}{2}\right)\left(\frac{n}{2}+1\right)}{2}$$

$$= \frac{n(n+2)(n+1)}{12} - \frac{n(n+2)}{8}$$

$$= \frac{n(n+2)(2n+2-3)}{24}$$

$$= \frac{n(n+2)(2n-1)}{24}.$$

Step 3

For all 'point down' triangles when n is odd we want $T_2 + T_4 + T_6 + \cdots + T_{2i}$ for $i = 1,2,3,\dots,\dfrac{n-1}{2}$.

$$\sum_{i=1}^{(n-1)/2} T_{2i} = \frac{1}{2}\sum_{i=1}^{(n-1)/2} 2i(2i+1) = \sum_{i=1}^{(n-1)/2}(2i^2+i)$$

$$= \frac{2\left(\frac{n-1}{2}\right)\left(\frac{n-1}{2}+1\right)\left(2\cdot\frac{n-1}{2}+1\right)}{6} + \frac{\left(\frac{n-1}{2}\right)\left(\frac{n-1}{2}+1\right)}{2}$$

$$= \frac{(n-1)(n+1)n}{12} + \frac{(n-1)(n+1)}{8}$$

$$= \frac{(n-1)(n+1)(2n+3)}{24}.$$

Steps 4 and 5
The intermediate steps are omitted here but appear in the original article (Eds.).

Combining all 'up' and 'down' triangles (i.e. adding the respective expressions) for *n* even gives

$$\frac{n(n+2)(2n+1)}{8}.$$

Combining all 'up' and 'down' triangles for *n* odd gives

$$\frac{(n+1)(2n^2+3n-1)}{8}.$$

Finally, we test these formulas against the results in Table 5:

$$n = 8: \quad \frac{8(8+2)(2\times 8+1)}{8} = 10 \times 17 = 170.$$

$$n = 7: \quad \frac{(7+1)(2\times 49 + 3\times 7 - 1)}{8} = 118.$$

Checks!

4.4 'Parallelograms of an $n \times n$ rhombus: A geometric counting problem', by David R Duncan and Bonnie H Litwiller

Vol. 8, no. 5 (November 1979), pp 26-29

A frequently encountered counting exercise in geometry asks for the number of rectangles that are contained on a checkerboard of a given size. The answer to this query yields

$$1^3 + 2^3 + 3^3 + \cdots + n^3 \, or \, (1 + 2 + 3 + \cdots + n)^2$$

rectangles for a checkerboard of size $n \times n$. For example, an 8×8 checkerboard contains $(1 + 2 + 3 + \cdots + 8)^2$ or 1296 rectangles. One of the authors investigated this question in a geometry class. The group had earlier worked with isometric dot paper. One alert student asked if the counting of rectangles could be accomplished on the isometric paper. After considerable discussion, the class decided that, because of the construction of the graph paper, parallelograms on a rhombus would be counted rather than rectangles on a square checkerboard. After much random counting the class decided that there were three types of parallelograms to be counted. Each type is illustrated on a 4×4 rhombus in Figures 1-3.

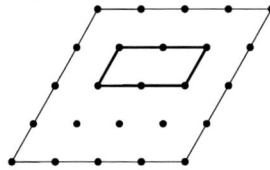

Fig. 1: Type I

Type I Parallelograms that have two sides horizontal and two sides of positive slope. Call this a 2×1 parallelogram since it has two units on the horizontal base and one unit along a side.

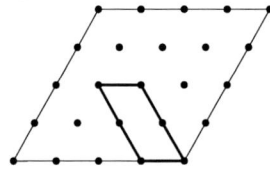

Fig. 2: Type II

Type II Parallelograms that have two sides horizontal and two sides of negative slope. Call this a 1×2 parallelogram since it has one unit on the horizontal base and two units along a side.

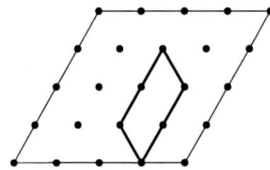

Fig. 3: Type III

Type III Parallelograms that have no horizontal sides. Call this a 2×1 parallelogram since its sides of positive slope are two units long and its sides of negative slope are one unit long. A 1×1 rhombus has only one 1×1 parallelogram - itself. This parallelogram is of Type I. A 2×2 rhombus has 13 parallelograms: 9 are of Type I (Fig. 4), 2 of Type II (Fig. 5), 2 of Type III (Fig. 6).

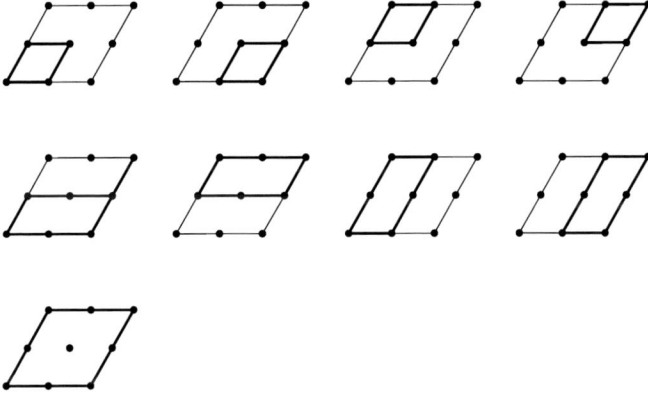

Fig. 4

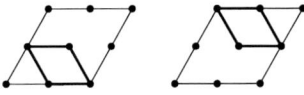

Fig. 5

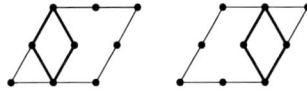

Fig. 6

Tables 1-6 report the total numbers of all types of parallelograms of all dimensions for rhombi of size $1 \times 1, 2 \times 2, ... 6 \times 6$ as generated by the class by actually drawing the figures.

1 by 1 rhombus			
Type I	Type II	Type III	Sum
1×1: 1	0	0	1

Table 1

2 by 2 rhombus			
Type I	Type II	Type III	Sum
1×1: 4 1×2: 2 2×1: 2 2×2: 1 9	1×1: 2	1×1: 2	13

Table 2

3 by 3 rhombus			
Type I	Type II	Type III	Sum
1 × 1: 9	1 × 1: 6	1 × 1: 6	58
1 × 2: 6	1 × 2: 2	1 × 2: 2	
1 × 3: 3	2 × 1: 3	2 × 1: 3	
2 × 1: 6	11	11	
2 × 2: 4			
2 × 3: 2			
3 × 1: 3			
3 × 2: 2			
3 × 3: 1			
36			

Table 3

4 by 4 rhombus			
Type I	Type II	Type III	Sum
1 × 1: 16	1 × 1: 12	1 × 1: 12	170
1 × 2: 12	1 × 2: 6	1 × 2: 6	
1 × 3: 8	1 × 3: 2	1 × 3: 2	
1 × 4: 4	2 × 1: 8	2 × 1: 8	
2 × 1: 12	2 × 2: 3	2 × 2: 3	
2 × 2: 9	3 × 1: 4	3 × 1: 4	
2 × 3: 6	35	35	
2 × 4: 3			
3 × 1: 8			
3 × 2: 6			
3 × 3: 4			
3 × 4: 2			
4 × 1: 4			
4 × 2: 3			
4 × 3: 2			
4 × 4: 1			
100			

Table 4

5 by 15 rhombus			
Type I	Type II	Type III	Sum
1 × 1: 25	1 × 1: 20	1 × 1: 20	395
1 × 2: 20	1 × 2: 12	1 × 2: 12	
1 × 3: 15	1 × 3: 6	1 × 3: 6	
1 × 4: 10	1 × 4: 2	1 × 4: 2	
1 × 5: 5	2 × 1: 15	2 × 1: 15	
2 × 1: 20	2 × 2: 8	2 × 2: 8	
2 × 2: 16	2 × 3: 3	2 × 3: 3	
2 × 3: 12	3 × 1: 10	3 × 1: 10	
2 × 4: 8	3 × 2: 4	3 × 2: 4	
2 × 5: 4	4 × 1: 5	4 × 1: 5	
3 × 1: 15	85	85	
3 × 2: 12			
3 × 3: 9			
3 × 4: 6			
3 × 5: 3			
4 × 1: 10			
4 × 2: 8			
4 × 3: 6			
4 × 4: 4			
4 × 5: 2			
5 × 1: 5			
5 × 2: 4			
5 × 3: 3			
5 × 4: 2			
5 × 5: 1			
225			

Table 5

6 by 6 rhombus			
Type I	Type II	Type III	Sum
1 × 1: 36	1 × 1: 30	1 × 1: 30	791
1 × 2: 30	1 × 2: 20	1 × 2: 20	
1 × 3: 24	1 × 3: 12	1 × 3: 12	
1 × 4: 18	1 × 4: 6	1 × 4: 6	
1 × 5: 12	1 × 5: 2	1 × 5: 2	
1 × 6: 6	2 × 1: 24	2 × 1: 24	
2 × 1: 30	2 × 2: 15	2 × 2: 15	
2 × 2: 25	2 × 3: 8	2 × 3: 8	
2 × 3: 20	2 × 4: 3	2 × 4: 3	
2 × 4: 15	3 × 1: 18	3 × 1: 18	
2 × 5: 10	3 × 2: 10	3 × 2: 10	
2 × 6: 5	3 × 3: 4	3 × 3: 4	
3 × 1: 24	4 × 1: 12	4 × 1: 12	
3 × 2: 20	4 × 2: 5	4 × 2: 5	
3 × 3: 16	<u>5 × 1: 6</u>	<u>5 × 1: 6</u>	
3 × 4: 12	175	175	
3 × 5: 8			
3 × 6: 4			
4 × 1: 18			
4 × 2: 15			
4 × 3: 12			
4 × 4: 9			
4 × 5: 6			
4 × 6: 3			
5 × 1: 12			
5 × 2: 10			
5 × 3: 8			
5 × 4: 6			
5 × 5: 4			
5 × 6: 2			
6 × 1: 6			
6 × 2: 5			
6 × 3: 4			
6 × 4: 3			
6 × 5: 2			
<u>6 × 6: 1</u>			
441			

Table 6

Size of rhombus	No. of parallelograms	Differences			
1 × 1	1				
2 × 2	13	12	33		
3 × 3	58	45	67	34	
4 × 4	170	112	113	46	12
5 × 5	395	225	171	58	12
6 × 6	791	396			

Table 7

An examination of Tables 1-6 led the group to make certain conjectures. These are most simply stated using the following notation: $n(w, x, y, z)$ means the number of parallelograms on a $w \times w$ rhombus of type x which have size $y \times z$. For example, $n(5, 1, 3, 4) = 6$ because a 5×5 rhombus contains six parallelograms of Type I with dimension 3×4 (Table 5).

Conjecture 1: $n(a, 1, b, c) = n(a, 1, c, b)$. The number of $b \times c$ Type I parallelograms is the same as the number of $c \times b$ Type I parallelograms on a given rhombus – a type of commutativity.

Conjecture 2: $n(a, 1, 1, 1) = a^2$, e.g. a 6×6 rhombus contains 36 1×1 parallelograms of Type I.

Conjecture 3: $n(a, 1, b, 1) = a^2 - (b - 1)a$. For each unit increase in b, the number of parallelograms decreases by a. $n(6, 1, 3, 1) = 6^2 - (3 - 1)6 = 24$. A perusal of Table 6 reveals that there are, in fact, 24 parallelograms of dimension 3×1 of Type I.

Conjecture 4: $n(a, 1, b, c) = n(a, 1, b, 1)$
$$= [(a - b + 1)(c - 1)]$$
$$= a^2 - (b - 1)a - [(a - b + 1)(c - 1)].$$
This is an especially valuable result since it gives a closed formula for the number of parallelograms of Type I of any given size on a specific rhombus. For example, on the 6×6 rhombus there are **24** 3×1 parallelograms of Type I, **20** 3×2, **16** 3×3, **12** 3×4, **8** 3×5, **4** 3×6. Note that $4 = a - b + 1 = 6 - 3 + 1$ was successively subtracted.

Conjecture 5:

$$\sum_{1 \leq b, c \leq a} n(a, 1, b, c) = 1^3 + 2^3 + 3^3 + \cdots + a^3 = (1 + 2 + 3 + \cdots + a)^2.$$

The numbers of parallelograms of Type I on rhombi of size $1 \times 1, 2 \times 2, \ldots 6 \times 6$ are respectively $1, 9, 36, 100, 225$ and 441.

Conjecture 6: $n(a, 2, b, 1) = n(a, 1, b + 1, 1)$. This relates Type I and Type II parallelograms. $n(6, 2, 3, 1) = n(6, 1, 4, 1)$. On a 6×6 rhombus there is the same number of 3 × 1 parallelograms of Type II as there of 4 × 1 parallelograms of Type I.

Conjecture 7: $n(a, 2, b, c) = n(a - 1, 2, b, c - 1)$. This enables one to count many of the parallelograms of Type II on a given rhombus by referring to a rhombus one size smaller. The numbers of 2 × 2, 2 × 3, 2 × 4 parallelograms of Type II on a 6 × 6 rhombus are respectively 15, 8, and 3. This is the same as the number of 2 × 1, 2 × 2, 2 × 3 parallelograms of Type II on a 5 × 5 rhombus.

Conjecture 8: $n(a, 2, b, c) = n(a, 3, b, c)$. Surprisingly, Types II and III have identical numbers of parallelograms of any given size on a given rhombus.

Conjecture 9: If the total number of parallelograms of all 28 types are listed for rhombi of increasing size and if consecutive differences are found, the fourth differences are all the constant 12 (Table 7).

If Table 7 were extended, the 7 × 7 rhombus would contain 1428 parallelograms; the 8 × 8 rhombus would contain 2388. The class then verified that these sums were correct.

Challenges for your class

- Verify each of the conjectures deductively.
- Find and verify other conjectures.
- A 38 × 38 rhombus is the smallest rhombus that would contain at least a million parallelograms. Calculate the exact number of parallelograms that it would contain.
- How many parallelograms would be contained on a rectangular $n \times n$ checkerboard?

4.5 'Pinboard isoperiminoes' by Erick Gooding
Vol. 16, no. 4 (September 1987), pp 20-22; extract

I would be surprised if I were alone in my view that pupils are often introduced far too quickly to the algorithm of multiplication for area. Surely a more fundamental idea is the composition and decomposition of area and the summing of parts. Perhaps what is needed is a collection of activities emphasising these ideas. I have often used tangrams to do this (and incidentally introduce some simple algebra) but other materials are needed to help generalisation. Recently a confluence of influences (among whom I gratefully acknowledge the three "G"s – Gardner, Golomb and Giles) has prompted me to try the activities described below. I offer them as starters for exploration. Symmetry is involved as an incidental, since it makes the puzzle aspect more viable, but the insistence on symmetry could be relaxed to make the idea more easily accessible.

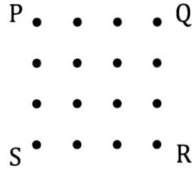

Fig. 1

1. The sixteen-pin board (Fig. 1) has a perimeter PQRS of 12 units. Find as many other shapes as you can on this pinboard that have a perimeter of 12 units. Record your results on squared paper. (Of course a loop of string or cotton is a valuable "tool" in the classroom.)

2. If we rule out disjointed shapes such as Figure 2, since soon we shall cut out a set of the shapes from card, the final set of shapes form a specific subset of the polyominoes. One covers 8 squares, some 7, some 6 and some 5 (see Figure 3.) Make a set from thin card. Notice that we do not need any mirror images since we can turn the cards over. (Obviously, discussion of whether shape are 'different' should have been engendered by Activity 1.)

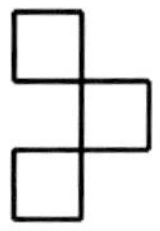

Fig. 2

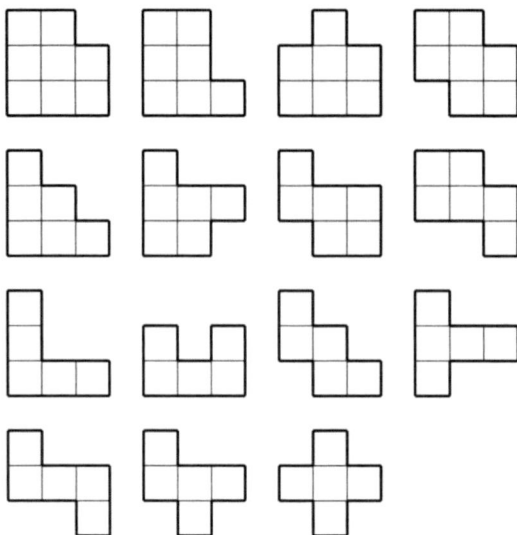

Fig. 3

Now by placing two or more of the pieces side by side so that parts of their edges meet, we can make outlines with various areas and perimeters. For example Figure 4a has area 13, perimeter 16, and Figure 4b has area 19, perimeter 20. Each can be made by various different combinations of pieces. Pupils can be presented with outlines of this sort as puzzles and can then make up others to set each other, with discussion and recording of the resulting areas and perimeters.

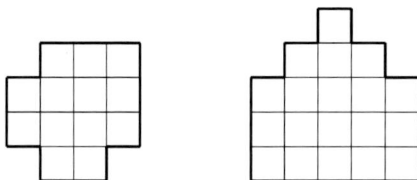

Figs. 4a (left) and 4b

3. Many of the pieces are symmetrical. Can you use combinations of the ones which are NOT to make symmetrical outlines with areas 11, 13, 17, 18, 24 and 29 units? Can you find other symmetrical outlines made from the asymmetric pieces? Work out the perimeter of each outline you find.

4. Using any of the pieces of Figure 3, how many symmetrical outlines can you make that cover 10 squares? What are their perimeters?

5. Again using any of the pieces, how many different symmetrical outlines can you make that have a perimeter of 20? Can you manage every possible area between 10 and 25 inclusive. (Are any other areas possible?)

6. Try to fit the complete set of 15 pieces into a rectangle.

4.6 'The number of rectangular prisms in a cube'
by Chris Pritchard
Vol. 45, no. 5 (November 2016), pp 33-35

Here we consider a 3-D analogue of the investigation of the number of oblongs on a chessboard: how many cuboids are there in a cube and how many of those cuboids are rectangular prisms, i.e. not cubes?

In the general case we have a cube of side n, constructed of n^3 unit cubes. But let us start as small as possible with $n = 1$ and build up some results. In this case there is just one cuboid and it is also a cube.

If $n = 2$, there are cuboids of four shapes:
- $1 \times 1 \times 1$ cube
- $1 \times 1 \times 2$ cuboid
- $1 \times 2 \times 2$ cuboid
- $2 \times 2 \times 2$ cube

Can you see that there are 8, 12, 6 and 1 respectively of these shapes; total = 27?

Now as we look at the $n = 3$ case the complexity increases, so that visualizing the possibilities becomes a demanding and rewarding challenge. A small part of meeting that challenge will be to count the number of $1 \times 1 \times 2$ cuboids. Though it's not the only way of doing it, we might be prompted by Euler's formula, and think in terms of vertices, edges and faces. In the top row of Figure 1 below, we picture three ways of placing such a cuboid in a corner of the cube. The second row shows the two ways in which the cuboid could be associated with a particular edge. The third row shows the single way in which the cuboid could be associated with a particular face.

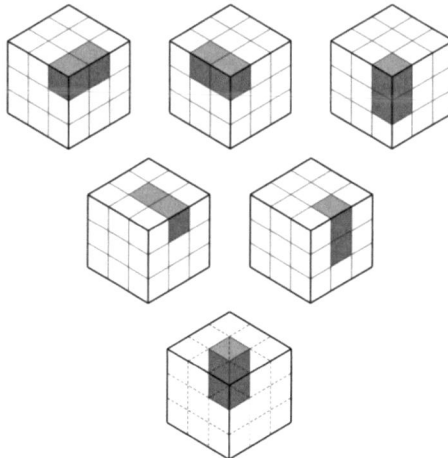

Fig. 1

Since a cube has 8 vertices, 12 edges and 6 faces, the total number of ways of arranging a $1 \times 1 \times 2$ cuboid in a $3 \times 3 \times 3$ cube is $8 \times 3 + 12 \times 2 + 6 \times 1 = 54$.

Alternatively, we could picture the $1 \times 1 \times 2$ cuboids according to the scheme in Figure 2. Here we initially focus on say the top layer. There are 6 positions in which the $1 \times 1 \times 2$ cuboid can be placed so that a 1×2 face shows on top. We can imagine sliding the cuboid from the first position to the second to the third (upper row of Fig. 2), and all over again a second time (lower row). But we can also imagine the $1 \times 1 \times 2$ cuboid occupying corresponding positions in the middle layer and again in the bottom layer. And then again we could rotate the cube into another two positions and get a further 18 results each time.

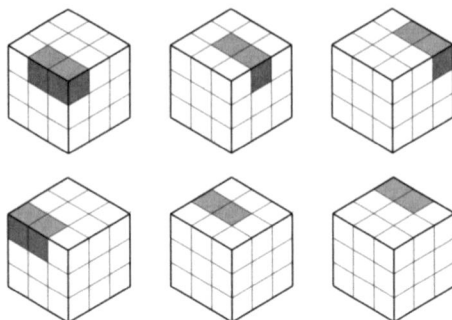

Fig. 2

In total, the number of $1 \times 1 \times 2$ cuboids in a cube of side 3, is $6 \times 3 \times 3 = 54$.

And there may be other ways of visualizing and counting the cuboids which will emerge when we try this in class. If that happens, all well and good!

The great challenge is to go through the process for all possible cuboids lurking in the cube. It's a big task which could reasonably be divided up around the class. A hands-on approach using blocks might be essential for some members of the class and it would certainly be a good idea to have them available for all pupils initially. On the other hand, we want to develop the ability to imagine, to visualize, so those pupils who can make progress without the blocks should be encouraged to do so. And if all goes well, the results in the Table 1 below should emerge.

Dimensions	f	Dimensions	f	Dimensions	f
$1 \times 1 \times 1$	**27**				
$1 \times 1 \times 2$	54				
$1 \times 1 \times 3$	27				
$1 \times 2 \times 2$	36	$2 \times 2 \times 2$	**8**		
$1 \times 2 \times 3$	36	$2 \times 2 \times 3$	12		
$1 \times 3 \times 3$	9	$2 \times 3 \times 3$	6	$3 \times 3 \times 3$	**1**

Table 1: number of cuboids in a cube of side 3 (total = 216)

And to crank up the complexity even further, to provide a challenge for those who are really good at visualizing, the number of cuboids of all sizes in a cube of side 4 could be sought. The results are as in Table 2 and some patterns begin to emerge. Have we seen the 8, 12, 6 and 1 before?

Dimensions	f	Dimensions	f	Dimensions	f	Dimensions	f
$1 \times 1 \times 1$	**64**						
$1 \times 1 \times 2$	144						
$1 \times 1 \times 3$	96						
$1 \times 1 \times 4$	48						
$1 \times 2 \times 2$	108	$2 \times 2 \times 2$	**27**				
$1 \times 2 \times 3$	144	$2 \times 2 \times 3$	54				
$1 \times 2 \times 4$	72	$2 \times 2 \times 4$	27				
$1 \times 3 \times 3$	48	$2 \times 3 \times 3$	36	$3 \times 3 \times 3$	**8**		
$1 \times 3 \times 4$	48	$2 \times 3 \times 4$	36	$3 \times 3 \times 4$	12		
$1 \times 4 \times 4$	12	$2 \times 4 \times 4$	9	$3 \times 4 \times 4$	6	$4 \times 4 \times 4$	**1**

Table 2: number of cuboids in a cube of side 4 (total = 1000)

Tables 1 and 2 have been positioned in such a way that a particular feature is apparent. It is the embedding or nesting of the frequency columns of Table 1 in Table 2. Why does that happen?

Rectangles in a square and cuboids in a cube

In the previous piece on the total number of squares, rectangles and oblongs in a square, we found that the formulas for the number of rectangles in a square of side n and the number of squares in a square of side n are, T_r and T_s, where

$$T_r = \left(\sum_{r=1}^{n} r \right)^2 = \left[\frac{1}{2} n(n+1) \right]^2 = \frac{1}{4} n^2 (n+1)^2,$$

$$T_s = \sum_{r=1}^{n} r^2 = \frac{1}{6} n(n+1)(2n+1).$$

The formula for the number of oblongs (non-square rectangles), T_b, was found by subtraction to be

$$T_b = \frac{1}{12} (n-1)n(n+1)(3n+2).$$

Do the formulas in the 3D analogue bear any structural resemblance?

The easiest place to start is with the cubes inside a cube of side n. In Tables 1 and 2 the frequencies or number of occurrences have been shown in bold and transferred into Table 3.

n	Occurrences	Totals
1	1	1
2	$1 + 8$	9
3	$1 + 8 + 27$	36
4	$1 + 8 + 27 + 64$	100

Table 3: number of cubes in a cube

Clearly, regardless of the size of the cube under consideration, we are seeing a 'sum of cubes' for the number of component cubes, and we know that a sum of cubes is itself a perfect square. And it's quite easy to explain using a model. So, labelling the number of component cubes T_c, the general formula is:

$$T_c = \sum_{r=1}^{n} r^3 = \frac{1}{4}n^2(n+1)^2 = \left(\sum_{r=1}^{n} r\right)^2.$$

We have already noted that there is just 1 cuboid in a cube of side 1 and there are 27 cuboids in a cube of side 2. From Tables 1 and 2, we know that there are 216 cuboids in a cube of side 3 and 1000 cuboids in a cube of side 4. And all four values are perfect cubes themselves:

$$1 = 1^3, 27 = 3^3, 216 = 6^3, 1000 = 10^3.$$

We can say more. They are the cubes of successive triangular numbers. In fact, the total number of cuboids, T_d, in a cube of side n is given by the formula:

$$T_d = \left(\sum_{r=1}^{n} r\right)^3.$$

This is a perfect analogue of the squares in a square scenario. Of course we have only prima facie evidence for this statement, and that certainly won't sway a jury of mathematicians.

Consequently, the formula for the number of rectangular prisms, T_p, in a cube of side n is:

$$T_p = \left(\sum_{r=1}^{n} r\right)^3 - \left(\sum_{r=1}^{n} r\right)^2 = \Delta_n^2(\Delta_n - 1),$$

where Δ_n denotes the nth triangular number. This factorises quite nicely into

$$T_p = \frac{1}{8}(n-1)n^2(n+1)^2(n+2),$$

which has a symmetrical run of central factors, mirroring in that respect the formula for the number of oblongs in a cube alluded to earlier,

$$T_b = \frac{1}{12}(n-1)n(n+1)(3n+2)$$

and perhaps begs the question whether T_b should be written as

$$T_b = \frac{1}{4}(n-1)n(n+1)\left(n+\frac{2}{3}\right).$$

So if we imagine that a Rubik's cube has a full set of unit cubes within it (rather than the mechanism at its core), the total number of rectangular prisms it contains is

$$\frac{1}{8} \times 2 \times 3^2 \times 4^2 \times 5 = 180$$

and this is fully in accord with the values reached earlier, i.e. $216 - 36$.

160

4.7 Michael Fox provides a 'Proof' of the formula for the number of cuboids in a cube'
Vol. 45, no. 5 (November 2016), p 35

Chris Pritchard conjectures in the article above that the sum of terms that gives T_d, i.e. the total of the results in Table 2, is the cube of a triangular number. It turns out that the conjecture is true for every cube having an edge length which is a positive integer, but we still need to prove it.

Consider a cube of side n, where n is a natural number; and place it with one vertex at the origin, and the three edges that meet there being along the x, y and z axes. Now take a cuboid with edges a, b and c, natural numbers that do not exceed n. Place it inside the cube with one vertex at $(0, 0, 0)$ and the opposite vertex at (a, b, c), its edges being parallel to the axes. (Regard this as distinct from cuboids with edges b, c, a; or c, a, b, etc.) We can translate it in the x direction by 1, or 2, or 3, ... or $n + 1 - a$ units; or in the y direction by 1, 2, 3, ... or $n + 1 - b$ units; or in the z direction by 1, 2, 3, ... or $n + 1 - c$ units.

Let $\alpha = n + 1 - a$, $\beta = n + 1 - b$, and $\gamma = n + 1 - c$. (Then $a = n + 1 - \alpha$, etc.; so as each of a, b and c range through all the values from 1 to n, so do α, β, γ, but in a different order.) By translation we obtain $\alpha\beta\gamma$ cuboids with edges (a, b, c).

We do this for all ordered triples a, b, c; and to each corresponds an ordered triple of independent translations α, β, γ, giving $\alpha\beta\gamma$ cuboids. The total number of cuboids that we can obtain from a cube of side n is therefore $\Sigma\alpha\beta\gamma$, where α, β, γ independently take all integer values from 1 to n. Hence we have

$$
\begin{aligned}
(1 \times 1 &+ 1 \times 2 + \cdots + 1 \times n + 2 \times 2 + \cdots + 2 \times n + \cdots n \times 1 + \cdots + n \times n) \\
&\times (1 + 2 + \cdots + n) \\
= (1 \times 1 &+ 1 \times 2 + \cdots + 2 \times 2 + \cdots 2 \times n + \cdots + n \times 1 + \cdots + n \times n) \\
&\times (1 + 2 + \cdots + n) \\
= (1 + 2 &+ \cdots + n)(1 + 2 + \cdots + n)(1 + 2 + \cdots + n) \\
= (1 + 2 &+ \cdots + n)^3 \\
= n^3 &(n + 1)^3 / 8.
\end{aligned}
$$

The 2-D version of this gives the known results for rectangles in a square; and, trivially, the 1-D version gives the result for line-segments on a line of given integer length.

4.8 'A square's rectangles equals a cube's cubes' by Gordon Haigh

Vol. 22, no. 1 (January 1983), pp 32-33

The number of rectangles in an $n \times n$ square is the same as the number of cubes in an $n \times n \times n$ cube. Surely there must be a simple link to explain this unlikely fact? I've looked hard for a mapping that will associate a particular rectangle with a particular cube. I've nearly done it, but not quite! Can anyone improve on my effort?

A square's rectangles

That the number of rectangles in an $n \times n$ square is $(1 + 2 + 3 + 4 + \ldots + n)^2$ can be seen by noting that each one can be represented by the projection of two of its adjacent sides onto two adjacent sides of the square: any two segments on adjacent sides of the square define a rectangle. On one side there are n segments of length 1, $n - 1$ of length 2, ... and one of length n and so $(1 + 2 + 3 + \cdots + n)$ altogether. These, combined with the same on the other side, give the result.

A cube's cubes

That the number of cubes in an $n \times n \times n$ cube is $1^3 + 2^3 + 3^3 + \cdots + n^3$ can be seen by counting them according to size: there are n^3 of side 1, $(n - 1)^3$ of side 2 ... That a square's rectangles equals a cube's cubes is the surprise that motivates the search for a simple explanation.

On reflection

Consider a 3×3 square and represent all of the rectangles in it by their bottom left hand corners. Moreover, segregate them according to shape and orientation:

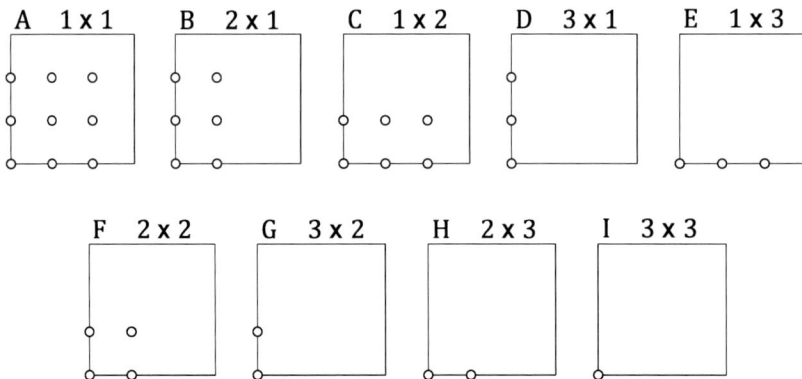

Fig. 1: The bottom left-hand corners of rectangles in a 3×3 square

Now represent all of the cubes in a $3 \times 3 \times 3$ cube by their bottom left vertices that sit in the planes sliced as shown:

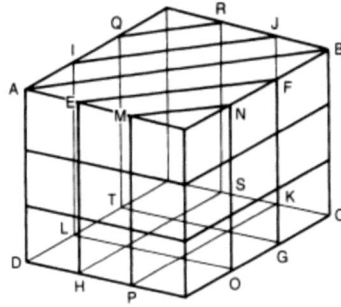

Fig. 2

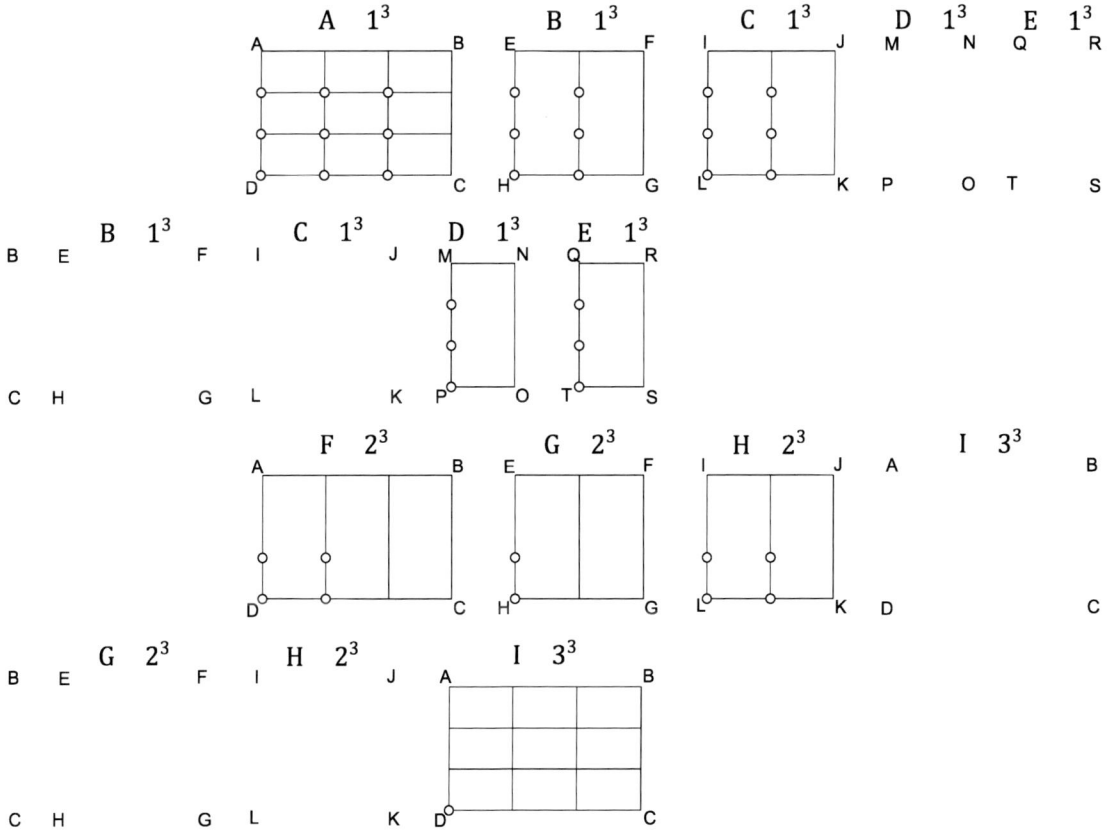

Fig. 3: The positions of the corners closest to the 'origin' of cubes in a
$3 \times 3 \times 3$ cube when sliced as in Fig. 2

Comparison of the figures shows the 1:1 correspondence. In fact it shows a tighter one: all rectangles of minimum length 1 in a $n \times n$ square with all cubes of side length 1 in an $n \times n \times n$ cube. The general case is similar. If only I didn't have to reflect the C, E and H diagrams to get an exact match, I could offer a formula which would tie it up neatly. Can it be done?

5 Number Bases and the Fibonacci Series

5.0 Introduction

In this chapter we include articles in which number bases and Fibonacci series are used as models in investigation. The articles by Paul Scott and Chris Pritchard, in particular illustrate that the knowledge and skills of binary numbers are needed to be able to explore the problems.

Number bases and Fibonacci series and their properties can themselves be introduced using structured investigations involving 'what if ...' type approaches. For example, for many years we developed the arithmetic of numbers in various bases, often creating elaborate stories to provide contexts for each scenario, such as a conversation between a two-toed sloth and a three-toed sloth on the number of berries they have picked. But probably the best context is that of cartoon characters, who, right back to Mickey Mouse, have been drawn with four digits on each hand. Indeed, Simon Singh's 2013 book *The Simpsons and their Mathematical Secrets* makes use of the idea. If thumbs are ignored, pupils can count on their eight 'true fingers': 1,2,3,4,5,6,7,10. (In fact, despite having only eight digits, almost forcing them to calculate in octal, the Simpsons actually use base ten because God, as depicted in the series, has ten digits and his word counts.)

What about adding and multiplying in various bases? We can develop addition and multiplication tables and then proceed with simple 'sums'. What about prime numbers? Easy, use the multiplication table and factors. Such activities are not merely esoteric; they reinforce the basic ideas of our own base ten system too. They highlight why arithmetic is as it is, why basic operations such as multiplication work the way they do.

Once the Fibonacci series is defined (1, 1, 2, 3, 5, 8, 13, ... or more formally as $F_1 = F_2 = 1, F_{n+1} = F_n + F_{n-1}, n > 2$), then more structured investigations can be introduced. Students are fascinated to discover the limit of the ratio of successive numbers, $\frac{F_{n+1}}{F_n}$, the so-called 'golden ratio'! What happens if you create your own Fibonacci type series, say 1, 3, 4, 7, 11, ... ? What happens to the limit of these ratios then?

Students not only find such activities fun, they stand to learn a great deal more than by simply moving through standard problems and exercises, while also consolidating their 'curriculum mathematics'.

5.1 'Taming the dragon' by Paul Scott
Vol. 26, no. 1 (January 1997), pp 2-4

Fig. 1: The dragon curve

Some paper-folding

To construct a 'dragon curve', we lay a long thin strip of paper flat on the table, and fold it over, right over left. Unfolding it will now give a V shape. Next fold the strip back again as it was, and then fold the folded strip again, right over left. Unfolding the paper, we now get angles V V Λ in this order.

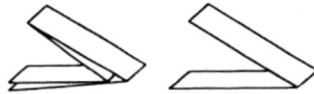

Exercise 1

Repeat this process several more times, and record your results. If we replace V by 1 and Λ by 0, we obtain in succession the sequences shown in Figure 2 – 'levels' of the *dragon sequence*. This sequence has a surprising number of interesting properties.

Let us denote the successive levels of the dragon sequence by $s_0, s_1, s_2, ...,$ as in Figure 2.

s_0	1
s_1	110
s_2	1101100
s_3	110110011100100
s_4	1101100111001001110110001100100

Fig. 2: Levels of the *dragon sequence*

The recursive structure of the sequences

Exercise 2

Look at the sequences you have obtained (Figure 2). What can you say about them? What would the next sequence be? More specifically,

- What is the middle digit in each row? Can you explain why these are all the same?
- Can you see that each row has a sort of reflective symmetry about this digit? Thinking of the paper-folding, can you explain why this occurs?
- How many digits are there in each row? Why do these numbers make us think of powers of 2?
- How many zeros are there in each row? How many ones?

Now it appears from Figure 2 (and is confirmed by the paper folding) that each sequence can be constructed from the one before in the following way. Let $\overline{s_k}$, denote the sequence obtained from s_k by reversing the order of the digits, and interchanging the zeros and ones. For example, $s_1 = 110$; reversing the order (011), and interchanging the zeros and ones gives $\overline{s_1} = 100$. Now we observe that

$$s_1 = s_0 1\overline{s_0}, s_2 = s_1 1\overline{s_1}, \dots \text{ and in general } s_{k+1} = s_k 1\overline{s_k}.$$

The presence of the central 1 each time, and the reflective structure of these sequences is easily explained from the paper folding. Looking at the simple cases leads us to conjecture that the number of digits in s_k is $2^{k+1} - 1$. This is certainly true for $k = 0, 1, 2, 3, 4$. An intuitive inductive argument might go like this: if s_k has $2^{k+1} - 1$ digits, then clearly so does $\overline{s_k}$. It follows that s_{k+1} has $(2^{k+1} - 1) + 1 + (2^{k+1} - 1) = 2 \cdot 2^{k+1} - 1 \equiv 2^{k+2} - 1$ digits, as expected. Similarly, by assuming that s_k has 2^k ones and $2^k - 1$ zeros, then by the reflective construction, $\overline{s_k}$ has $2^k - 1$ ones and 2^k zeros. Thus including the central 1, s_{k+1} has $2^k + 1 + 2^k - 1 = 2^{k+1}$ ones and, similarly, $2^{k+1} - 1$ zeros.

A formula for the digits in the sequence

Although the sequence structure we have obtained is interesting, it is still rather clumsy in practice. To find the digits at any given level, we have to construct each previous level of the sequence in turn. Might we not find a formula which gives the digit in the nth place of the sequence? Suppose we denote by $f(n)$ the value 0 or 1 corresponding to the nth fold (the nth digit from the left). We first obtain some data about these values by looking more closely at the paper-folding operation.

It is clear that the powers of 2 have an important role to play here: at some stage, each will occur in a centre-fold position. So we can place a one in each of these positions. The other terms of the sequence are then filled in successively as in Figure 3.

	1	2	3	4	5	6	7	8	9	10	11	12	13	14	15	16
s_1	1	1	0													
s_2	1	1	0	1	1	0	0									
s_3	1	1	0	1	1	0	0	1	1	1	0	0	1	0	0	
.	.	.	.	.	.	.	.	.	.	.	.	.	.	.	.	.

Note that in the top row, the bold font is used to indicate the positions of the centre folds (i.e. powers of 2) and in subsequent rows, for the **1** digits in s_1, s_2, s_3 ... about which reflection in the centre folds take place.

Exercise 3

We see that $f(n) = 1$ for $n = 1, 2, 4, 8, \ldots$ Can you conjecture values of $f(n)$ for other values of n? More specifically,

- What if $n = 1, 5, 9, 13, \ldots$?
- What if $n = 3, 7, 11, 15, \ldots$?
- What if $n = 1, 3, 5, 7, \ldots$?
- Can you see any relationship between $f(n)$ and $f(n/2)$ when n is even?

Here is a brief outline of some conclusions and why they are true. Reproducing the last row of the table above, we have

1	2	3	4	5	6	7	8	9	10	11	12	13	14	15	16	17	...
1	1	0	1	1	0	0	1	1	1	0	0	1	0	0	1	1	...

Because of the reflective property (reflecting successively in 2, 4, 8, 16, ,..) the values of $f(n)$ in the lower line have the repeating pattern

$$\bullet 110 \bullet 100 \bullet 110 \bullet 100 \bullet 110 \ldots$$

It follows that $f(1) = f(5) = f(9) = \ldots = 1$, that $f(3) = f(7) = f(11) = \cdots = 0$, and that $f(1), f(3), f(5), \ldots$ alternate between 1 and 0. Also if we consider the reflective property as it applies to the even values of n we obtain:

2	4	6	8	10	12	14	16	18	20	22	24	26	28	30	32	...
1	1	0	1	1	0	0	1	1	1	0	0	1	0	0	1	...

Comparing the lower lines of the last two tables we observe that the initial values are the same, and the method of formation of the sequence is the same. Hence the two rows of zeros and ones are identical. Hence the nth digit in the first row is the same as the nth digit in the second. That is, $f(n) = f(2n)$, as required.

Some further problems

There are many other interesting properties of the dragon sequence. Here are a few you might like to try to prove.

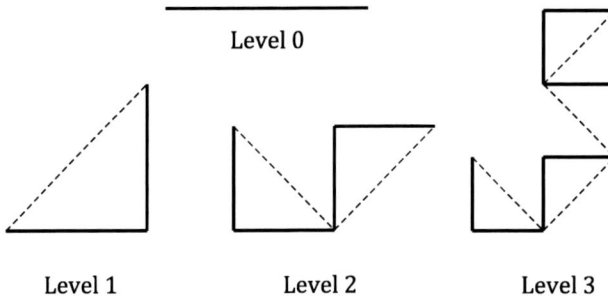

Fig. 3

- The sum of the digits of s_k is 2^k.
- For $k > 1$, s_k begins with 11 and ends with 00.

Let us define a *run* to be a sequence of identical digits. Then

- No run is longer than three digits.
- In s_k there are $2^{k-2} - 1$ runs of three repeated digits.
- In s_k there are $2^{k-1} + 1$ runs of two repeated digits.
- In s_k there are 2^{k-2} (runs of one digit) single digits.
- For s_k with $k > 1$, there is no case of a single one followed by a single zero.
- If we think of s_k as a number in base 2 arithmetic, then s_k is divisible by 3.
- If we think of s_k as a number in base 10 arithmetic, then s_k is divisible by 11.
- In general, if we think of s_k as a number in base n arithmetic, then s_k is divisible by $n + 1$.
- Only the middle digits of s_k and $\overline{s_k}$ differ.
- In each s_k, all previous s_j $(j < k)$ occur in order, separated by a zero or a one.

The dragon curve

The *dragon curve* is simply the shape formed by our repeatedly folded strip of paper, except that all the folds make angles of exactly 90°. The corresponding curves, in order of complexity, are shown in Figure 3. The dotted lines indicate the curve at the previous level, turned through 45°. The Level 14 curve is pictured in Figure 1.

5.2 William Gibbs on 'Paper weaving'
Extract from 'Paper patterns 4: Paper weaving', Vol. 19, no. 5 (November 1990), pp 16-19

In all countries and in all cultures, people have woven strips of leaf or bark or wool to create mats and baskets with intricate patterns and designs. This activity can be successfully simulated with strips of coloured paper and by varying the colour and order of the horizontal and vertical strips the fascinating variety of patterns created can be investigated.

Start by considering the simple weave in which each strand goes over and under alternately. What pattern will be created if the horizontal strips are of one colour, and the vertical strips of another? The result, alternating squares of each colour is not surprising.

However if both the horizontal and the vertical strips alternate in colour then the result is surprising.

It is satisfying first to predict the pattern and then to weave it. To make it easy to weave, cut strips of equal width from coloured paper, arrange the vertical strips in the pattern required and tape them along the top to the table or other firm surface. The crossing strips can then be woven in quickly and easily. (Pre-cut coloured gummed strips designed for making paper chains can be used). Here is another interesting pattern created using the simple weave. If the vertical strips are arranged in a repeating order, black, black, white and the horizontal strips are woven in the order black, white, white then a new tessellation emerges.

Many of the patterns woven traditionally display repeating patterns involving symmetry and tessellation. Here for example is the 'Star' pattern, or 'Shepherds' check', well known to weavers throughout the world.

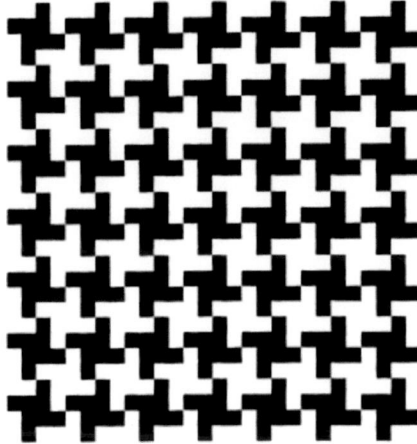

How is this tessellation created? It is an interesting challenge to try and deduce the weaving pattern used. In fact double strands of each colour alternate in both directions but although this result can be derived logically there is nothing like actually weaving it to confirm one's reasoning.

This and other patterns prompt the questions

- Can other tessellations be created
- Can the weavings needed be described mathematically?
- Can a simple algorithm be derived to determine the colour of any square in the pattern?

If we concentrate on patterns created using two colours then the order of the horizontal and vertical strips can be binary coded. For example the 'Star' pattern which had the strips arranged black, black, white, white can be coded as 1100 on both axes with 1 standing for black and 0 for white.

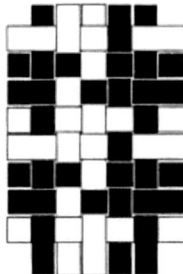

Here is another tessellation for which the code is 10010110 on both axes.

Looking at the structure of the code that created this tessellation notice that it has two properties. Firstly the code can be divided into two halves one of which is the complement of the other, 1001 and 0110. Secondly, the code has two lines of symmetry. As the code is repeated when making the pattern, all codes of 8 elements taken from this sequence are equivalent (10010110, 00101101, ... 01001011). These two codes are equivalent and both are symmetrical. Using these two ideas and creating sequences that have symmetry and complementary halves it is possible to create any number of tessellations.

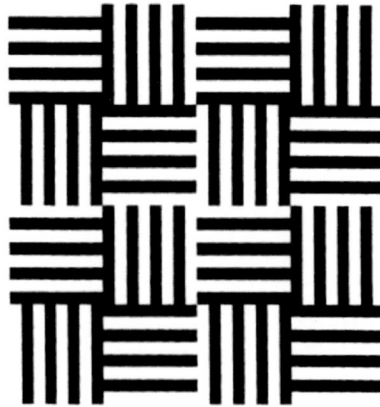

The pattern shown above is sometimes known as the 'Log Cabin' design. Try the binary equivalents of 43860, 699732 or 11187540 for even more elaborate designs of the same kind.

A simple algorithm for predicting the colour of any square in the pattern can be derived by considering the nature of the woven pattern. If the (1,1) square has the first vertical strip on top then so will (1,3), (1,5), (2,2), (3,5), (4,4) ... In general if $(x + y)$ is even then the vertical strip will be on top, if $(x + y)$ is odd then the horizontal strip will cover the vertical.

172

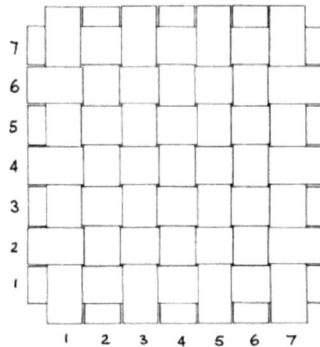

The colour of the strip will depend on the modular nature of the code being used. For example a pattern with strips arranged in the order black, black, white (i.e. with code 110) can be considered as modular 3. The coordinates now have to be converted to this mode to discover the correct colour. For example consider the square represented by (8,6); 8 + 6 = 14. The result is even so the vertical strip is on top. Mode 3 is 2. As the code is 110 the 2nd strip is "1". The square is thus black.

This analysis of the weave provides an excellent basis for creating a computer program which will generate weaving patterns and allow for the investigation of tessellations. To widen the investigation still further consider weaves with more than 2 colours. Here for example is another tessellation created using strips of three different colours arranged in sequence, as shown below.

5.3 'Binary bubbles' by Chris Pritchard

This inspiration for this article was Don Steward's 'Bubbles' in Vol. 16, no. 2 (March 1987), pp 42-43. It was published in our sister journal, SYMmetry+ 50 (Spring 2013), pp 3-4, 15.

Here are two bubbles drawn in two different ways, alongside each other in the first configuration and one inside the other in the second:

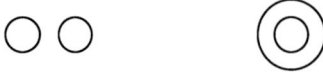

The sizes of the bubbles is unimportant, as are their positions as such, though the inside/outside concept is fundamental. And instead of circles simple loops are fine, and this means that when what follows is used in the classroom, lots of freehand, sketch diagrams can be generated quickly.

So, how many different configurations are there in the case of three bubbles?

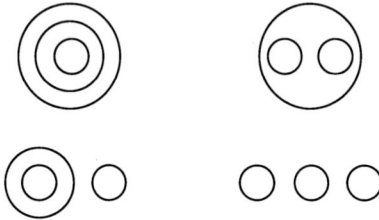

Of course, as the number of bubbles increases, it becomes trickier to find all the patterns. For four bubbles there are nine patterns, a 5-minute exercise for many pupils, with lots of opportunity to work together and share ideas and approaches (which might include adding an extra bubble judiciously to some of the four patterns above). But is there an efficient way of generating the patterns, albeit one that might dull the pleasure of pure exploration and invention?

Here's one possibility, which can be explained by taking this diagram

and cropping it at the top and bottom:

$$((\;)) (\;)$$

We get a sequence of brackets: (( )) ( ). The other three configurations of bubbles give, respectively:

Top left: ((( )))
Top right: (( )( ))
Bottom right: ( ) ( ) ( ).

We can take this coding further, using a 0 for an opening bracket and a 1 for a closing bracket. The 3-bubble patterns, bracket sequence and binary code is:

```
( ( ( ) ) )        ( ( ) ( ) )
  000111            001011
```

```
( ( ) ) ( )        ( ) ( ) ( )
  001101            010101
```

Since every sequence of brackets begins with an opening bracket and finishes with a closing bracket, perhaps they could be ignored, as could the first and last digits in the binary codes, which then become:

$$0011 \quad 0101$$
$$0110 \quad 1010$$

These four abbreviated codes each have two zeros and two 1s. (This seems perfectly reasonable, since every bracket, once opened, should be closed.) But there are two other binary codes consisting of two zeros and two 1s:

$$1001 \quad 1100$$

Working in reverse now, can bubble patterns relating to these abbreviated binary codes be generated? What would they look like?

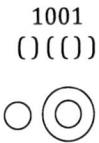

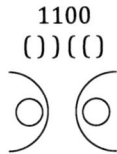

```
        1001                  1100
      ( ) ( ( ) )           ( ) ) ( ( )
```

These diagrams prompt the following questions:

1. For the first part of our discussion we would surely have considered the two bubble configurations below as one and the same, but is there a connection between their binary codes?

2. Can we incorporate the impossible or 'burst' bubble diagram, or perhaps 'infinite bubble' in our analysis?

To answer the first, we must use the full, rather than the abbreviated, binary codes and make a comparison:

```
( ( ) ) ( )        ( ) ( ( ) )
  001101            010011
```

The association is revealed by reversing the order of the binary digits and then taking complements (swapping zeros and 1s). If we carry out the same operations on the other three binary sequences, they do not change. Some sort of reflective property is apparent in the sequences and the bubble patterns.

If we are drawing our diagrams on a plain/plane sheet of paper the burst bubble should be discarded. But we can incorporate it if we think of it drawn on the curved surface of a cylinder, for example. Then, from an appropriate vantage point, it would look like:

We can see the relationship between the burst bubble and the button design above. It is cyclical in nature:

Button 011001 ↔ 001011 Burst

Before considering the four-bubble scenarios, some thought might be given to how many patterns we might be seeking. The abbreviated binary code would consist of 6 digits, three zeros and three 1s. There are $\binom{6}{3} = 20$ such patterns.

Code	Diagram	Code	Diagram
0 0 0 1 1 1 **A**		1 0 0 0 1 1 **K**	
0 0 1 0 1 1 **B**		1 0 0 1 0 1 **L**	
0 0 1 1 0 1 **C**		1 0 0 1 1 0 **M**	
0 0 1 1 1 0 **D**		1 0 1 0 0 1 **N**	
0 1 0 0 1 1 **E**		1 0 1 0 1 0 **O**	
0 1 0 1 0 1 **F**		1 0 1 1 0 0 **P**	

0 1 0 1 1 0 **G**		1 1 0 0 0 1 **Q**	
0 1 1 0 0 1 **H**		1 1 0 0 1 0 **R**	
0 1 1 0 1 0 **I**		1 1 0 1 0 0 **S**	
0 1 1 1 0 0 **J**		1 1 1 0 0 0 **T**	

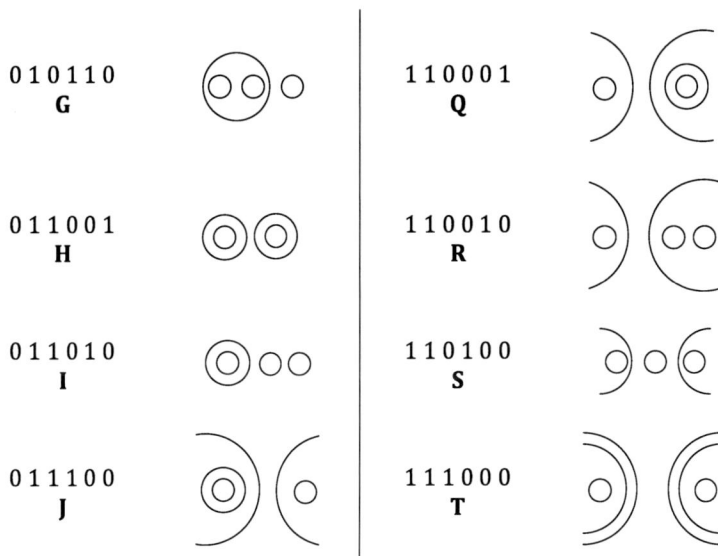

Extension: 'Exploring binary bubbles'

Extract taken from Presidential Essay 5 (August 2021), Mathematical Association website

In the first extension, we gave all 20 bubble diagrams in order of their binary code, and labelled **A** to **T** for convenience. Numerous observations might be made:

- Six bubbles are burst. They are **J, P, Q, R, S, T**. Now **S** and **T** are symmetrical, while the remaining four constitute two sets of twins: (**J, Q**), (**P, R**).
- Of the 14 valid bubbles, six are symmetrical (**A, B, F, H, M, O**) and there are three sets of twins, (**C, E**), (**D, K**) and (**G, L**).
- (**I, N**) is also a reflective pair but both are variants of the symmetrical **M** and should be discounted. Or to put it another way, **M, I** and **N** form a triplet.

So in total, there are nine different, valid bubble designs, represented by (for example) **A, B, C, D, F, G, H, M** and **O**.

Task: Five-bubble designs

(Homework task for individual pupils or as a group task in class.)

Find all the different five-bubble designs, identifying symmetrical bubbles, bubbles that have reflective twins and burst bubbles and make observations, frame conjectures and draw conclusions throughout.

Possible Solution

The number of possible bubbles is the number of sequences of eight digits, four 1s and four zeros,

$$\binom{8}{4} = 70.$$

They are shown in the table below. The 28 sequences highlighted in grey represent burst bubbles. They are those starting with two 1s or with three 1s in the first four or with four 1s in the first six (or those ending with two zeros). This leaves 42 binary sequences for which bubble arrangements are possible.

<u>00001111</u>	**00111100**	<u>01101001</u>	10011001	**11000101**
<u>00010111</u>	01000111	01101010	10011010	**11000110**
00011011	01001011	01101100	10011100	11001001
00011101	<u>*01001101*</u>	01110001	10100011	11001010
00011110	01001110	01110010	10100101	11001100
00100111	01010011	01110100	10100110	11010001
<u>00101011</u>	<u>01010101</u>	01111000	10101001	11010010
00101101	01010110	10000111	<u>10101010</u>	11010100
00101110	01011001	10001011	10101100	11011000
<u>00110011</u>	01011010	10001101	10110001	11100001
00110101	*01011100*	<u>10001110</u>	10110010	11100010
00110110	01100011	10010011	10110100	11100100
00111001	01100101	10010101	10111000	11101000
00111010	01100110	<u>10010110</u>	11000011	<u>11110000</u>

These valid sequences fall into four subsets. The first two subsets have 10 distinct members altogether (underlined); they represent symmetrical bubble diagrams. When looked at in a mirror or from behind they remain unchanged. But they are of two types. Six of them are of a simple, singlet form (upper two rows, below) and the remaining four are found in triplets (bottom row).

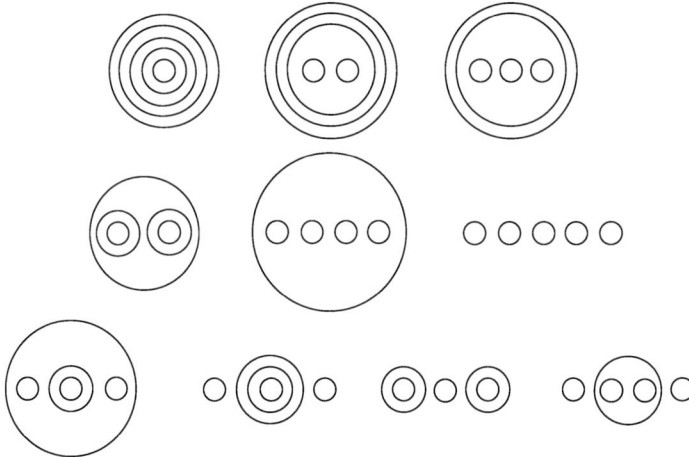

The remaining 32 binary sequences have a pairwise reflective property. There are 8 sets of twins (highlighted in bold font) and two sets of quadruplets (one set in italics and one set underlined).

00011011 **00100111,** **00011101,** **01000111** **00011110,** **10000111** **00101101,** **01001011**	**00101110,** **10001011** 00110101, 01010011 *00110110,* *10010011* **00111001,** **01100011**	00111010, 10100011 *01001110,* *10001101* **01010110,** **10010101** **01011001,** **01100101**	01011010, 10100101 01100110, 10011001 01101010, 10101001 10011010, 10100110

The first of each pair of these 8 sets of twins is drawn below.

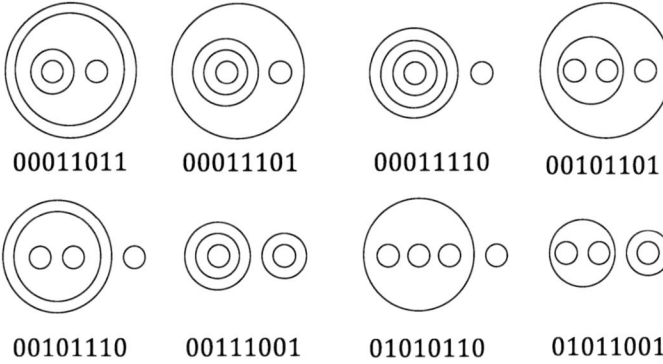

00011011   00011101   00011110   00101101

00101110   00111001   01010110   01011001

The first set of quadruplets is shown below.

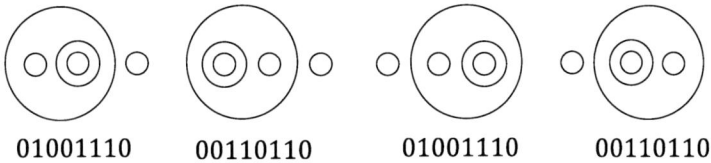

01001110   00110110   01001110   00110110

The other set is represented by:

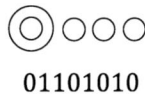

01101010

Each of the pairs with no highlighting are versions of one of the symmetrical designs already drawn. For example, 00111010 and its reflection, 10100011, belong to 10001110.

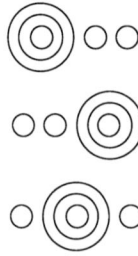

So there are four triplets (not four additional pairs).

So altogether, we have 6 symmetrical singlets, 4 symmetrical triplets, 8 sets of twins and 2 sets of quadruplets. That's 20 distinct, valid bubbles.

Note

Zoltán Retkes has drawn the Editors' attention to the connection between the above results and both Catalan and Narayana numbers. He comments:

> When you count the valid configurations you may classify the bracket sequences according to the number of nestings, e.g., if $n = 4$ then $(()(()))$ contains two nestings while $((()))$ contains one. The number of valid sequences with k nestings is given by the Narayana number $N(n, k)$ for n pairs of brackets. In this example $N(4,1) = 1$, $N(4,2) = 6$, $N(4,3) = 6$, $N(4,4) = 1$. The sum of these numbers is $1 + 6 + 6 + 1 = 14$, the fourth Catalan number. For $n = 5$ you would have $1 + 10 + 20 + 10 + 1 = 42$, the fifth Catalan number and so on.

The nth Catalan number and the nth Narayana number are given by the respective formulas,

$$C_n = \frac{1}{n+1}\binom{2n}{n} = \frac{(2n)!}{n!\,(n+1)!},$$

$$N(n, k) = \frac{1}{2}\binom{n}{k}\binom{n}{k-1}.$$

They are named after the Belgian mathematician, Eugène Charles Catalan (1814–1894) and the Canadian mathematician, Tadepalli Venkata Narayana (1930–1987), and they feature in or are closely connected with Pascal's triangle (attributed in some parts of the world to Omar Khayyam or Jia Xian rather than Blaise Pascal). So plenty of scope for further investigation!

5.4 'A balanced approach to number bases' by Chris Pritchard
Vol. 45, no. 3 (May 2016), pp 15-18

This article investigates two problems and some of the mathematics associated with the second of them. A number of tasks are given, as are their solutions.

Problem 1

We have a pan balance which we intend using to weigh some object. The object is placed on one pan while the **weights go only on the other pan**. The smallest weight is of 1 unit and there is only one such weight. All other weights are multiples of the basic unit weight and there is only one of each. Which weights should we choose to produce an efficient system of weighing?

Solution

The solution can be found by taking a systematic approach. The object weighs:

1 unit   we can weigh it using the basic unit weight

2 units   we have just one unit weight, so we need a new 2-unit weight

3 units   we can use the 2-unit and 1-unit weights together

4 units   we cannot make 4 using 2 and 1, so we adopt a new weight of 4 units

5 units   we use the 4 and 1 weights

6 units   we use the 4 and 2 weights

7 units   we use the 4 and 2 and 1 weights,   and so on.

Perhaps we lay this out using standard symbols:

$1 = 1$
$2 = 2$
$3 = 2 + 1$
$4 = 4$
$5 = 4 + 1$
$6 = 4 + 2$
$7 = 4 + 2 + 1,$

or perhaps we use a table:

		Available weights for measurement		
		4	**2**	**1**
	1	0	0	1
	2	0	1	0
	3	0	1	1
Weight of object	**4**	1	0	0
	5	1	0	1
	6	1	1	0
	7	1	1	1

and see immediately that we are using binary.

So if we wished to check the weight of a 25 unit sack, we could proceed by writing 25 as 11001_2 and placing the 16, 8 and 1 unit weights on the pan.

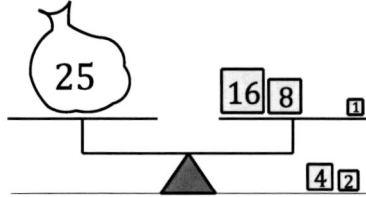

Problem 2

We have a pan balance which we intend using to weigh some object. The object is placed on one pan while the **weights may be placed on either pan**. Once more, the smallest weight is of 1 unit and there is only one such weight. All other weights are multiples of the basic unit weight and there is only one of each. Which weights should we choose now?

Solution

Again, let's be systematic. The object weighs:

1 unit   we can weigh it using the basic unit weight

2 units   we could adopt a 2-unit weight as before or else adopt a 3-unit weight and weigh an object of 2 units by balancing the object and the unit weight against the 3-unit weight

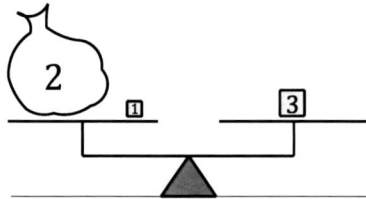

Under the second system, we would continue:

3 units   we can use the 3-unit weight on the opposing pan

4 units   we can use the 3 and 1 weights together on the opposing pan

5 units   now we introduce a 9 (= 3^2) weight, place it on the pan with the object, which also has the 3 and the 1 weights

6 units   the object and the 3 weight is balanced against the 9 weight

7 units   the object and the 3 weight is balanced against the 9 and 1 weights

and so on.

In simple arithmetic, we have:

1 = 1
2 = 3 – 1
3 = 3
4 = 3 + 1
5 = 9 – 3 – 1
6 = 9 – 3
7 = 9 – 3 + 1
8 = 9 – 1
9 = 9
10 = 9 + 1
11 = 9 + 3 – 1.

This is effectively base 3 arithmetic, known as ternary. But it is not *standard ternary*, which uses just the digits 0, 1 and 2 and in which we represent the numbers from 1 to 11 as:

$$1, 2, 10, 11, 12, 20, 21, 22, 100, 101, 102.$$

The version of ternary we have here is called *balanced ternary* or *signed ternary*. The former term presumably comes from the context of the 2-pan weighing exercise and the latter from the need to use negatives or signed numbers in amongst the positives. So let's have a looked at signed ternary and some of its weird but wonderful elementary arithmetic.

Balanced ternary uses the digits $\bar{1}$, 0 and 1, with $\bar{1}$ effectively meaning -1. We can get an idea of how the system would work by going back to the 2-pan balance and the example of weighing 7 units:

'7 units the object and the 3 weight is balanced against the 9 and 1 weights'

i.e. we need 9 positive, 3 negative and 1 positive and so write it as $1\bar{1}1$. Similarly, for 25 would need 27 positive, no 9, 3 negative and 1 positive, representing in terms of the scales, a 27 and a 1 on one side and the object and a 3 on the other. In balanced ternary:

	27	9	3	1
25	1	0	$\bar{1}$	1

Task 1

Generate the balanced ternary numbers equivalent to our standard numbers from 1 to 30. As a help, here are the first few:

1	1
2	$1\,\bar{1}$
3	$1\,0$
4	$1\,1$
5	$1\,\bar{1}\,\bar{1}$

The full list is given at the end of the article.

Just how easy is it to change between ternary and balanced ternary? From the list, we know that the balanced ternary for 7 is $1\,\bar{1}\,1$ and in standard ternary it is 21. Of course, there is no digit 2 in balanced ternary; it is replaced by $1\,\bar{1}$ and this leads directly to the representation we have for 7 in balanced ternary.

Task 2

(1) Change the following ternary numbers into balanced ternary:

$$102, 211, 10201.$$

(2) Change the following balanced ternary numbers into ternary:

$$1\,0\,\bar{1},\ 1\,\bar{1}\,1\,\bar{1},\ 1\,0\,0\,\bar{1}\,\bar{1}.$$

Addition

Adding in balanced ternary is especially easy once the rules are known, and it turns out the rules are easily generated. We are using just three symbols, and they are effectively -1, 0 and +1 in our system. The first seven cells of the addition table slot in immediately:

+	$\bar{1}$	0	1
$\bar{1}$		$\bar{1}$	0
0	$\bar{1}$	0	1
1	0	1	

The bottom right cell is for $1 + 1$, but we know that to be $1\,\bar{1}$. The top left cell is its negative, and to take a negative in balanced (or signed) ternary, just swap all signs: $-(1\,\bar{1}) = \bar{1}\,1$. Now we can complete the table.

+	$\bar{1}$	0	1
$\bar{1}$	$\bar{1}\,1$	$\bar{1}$	0
0	$\bar{1}$	0	1
1	0	1	$1\,\bar{1}$

Laying the calculations in the manner of long multiplication can be helpful at first but this scaffolding is not needed for long.

Example 1: $1\,\bar{1}\,\bar{1} + 1\,\bar{1}\,1$

$$
\begin{array}{r}
1\,\bar{1}\,\bar{1} \\
1\,\bar{1}\,1 \\
\hline
0 \\
\bar{1}\,1 \\
\underline{1\,\bar{1}} \\
1\,1\,0
\end{array}
$$

Example 2: $1\,\bar{1}\,0 + 1\,1\,\bar{1}$

$$
\begin{array}{r}
1\,\bar{1}\,0 \\
1\,1\,\bar{1} \\
\hline
1\,\bar{1}\,0\,\bar{1}
\end{array}
$$

Task 3

Carry out the following additions:
(1) $1\,\bar{1}\,0 + 1\,0\,1$
(2) $1\,0\,\bar{1}\,1 + 1\,0\,\bar{1}$
(3) $1\,\bar{1}\,\bar{1}\,0\,1 + 1\,0\,\bar{1}\,\bar{1}$

Subtraction

Task 4

In the manner of the addition table, create a subtraction table.

Example: $1\,\bar{1}\,0\,1 - 1\,\bar{1}\,0$

Using the subtraction table it is easy to get to this point:

$$
\begin{array}{r}
1\,\bar{1}\,0\,1 \\
1\,\bar{1}\,0 \\
1\,1\,1 \\
\hline
\bar{1}
\end{array}
$$

Having completed the calculation for the third column from the right, we have $\bar{1}\,1$. We put down the 1 and carry the $\bar{1}$ and act as if adding. The answer is 1 1 1.

Task 5

Carry out the following subtractions:

(1) $1\,1\,\bar{1}\,0 - 1\,0\,0$
(2) $1\,0\,\bar{1}\,1 - 1\,0\,\bar{1}$
(3) $1\,\bar{1}\,\bar{1}\,0\,1 - 1\,0\,\bar{1}\,\bar{1}$

Multiplication

Task 6

Create a multiplication table.

Example: $1\,0\,1 \times 1\,\bar{1}\,1$

$$
\begin{array}{r}
1\,0\,1 \\
1\,\bar{1}\,1 \\
\hline
1\,0\,1 \\
\bar{1}\,0\,\bar{1} \\
1\,0\,1 \\
\hline
1\,0\,\bar{1}\,\bar{1}\,1
\end{array}
$$

Task 7

Carry out the following calculations:

(1) $1\bar{1}0 \times 101$
(2) $10\bar{1}1 \times 1\bar{1}1$
(3) $1\bar{1}001 \times 10\bar{1}$
(4) $11\bar{1}^{2}$

I hope your pupils will enjoy an excursion into the arithmetic of balanced ternary. The answers to the questions are given below.

Answers

Task 1

1	1	16	$1\bar{1}\bar{1}1$
2	$1\bar{1}$	17	$1\bar{1}0\bar{1}$
3	10	18	$1\bar{1}00$
4	11	19	$1\bar{1}01$
5	$1\bar{1}\bar{1}$	20	$1\bar{1}1\bar{1}$
6	$1\bar{1}0$	21	$1\bar{1}10$
7	$1\bar{1}1$	22	$1\bar{1}11$
8	$10\bar{1}$	23	$10\bar{1}\bar{1}$
9	100	24	$10\bar{1}0$
10	101	25	$10\bar{1}1$
11	$11\bar{1}$	26	$100\bar{1}$
12	110	27	1000
13	111	28	1001
14	$1\bar{1}\bar{1}\bar{1}$	29	$101\bar{1}$
15	$1\bar{1}\bar{1}0$	30	1010

Task 2

(1) $11\bar{1}$, $1\bar{1}11$, $11\bar{1}01$

(2) $10\bar{1}$ is the number before 100, which is 22;
$1\bar{1}1\bar{1}$ may be split into two pairs of digits, each worth 2 in standard ternary; so the answer is 202;

(3) $100\bar{1}\bar{1}$ could be changed into standard ternary be counting back four (in our system) from 10000, i.e. 2222, 2221, 2220, **2212**.

Task 3 $1\bar{1}\bar{1}1$, $11\bar{1}0$, $10\bar{1}\bar{1}0$

Task 4

–	$\bar{1}$	0	1
$\bar{1}$	0	$\bar{1}$	$\bar{1}$ 1
0	1	0	$\bar{1}$
1	1 $\bar{1}$	1	0

Task 5   1 0 $\bar{1}$ 0, 1 $\bar{1}$ 0 $\bar{1}$, 1 0 $\bar{1}$ $\bar{1}$

Task 6

×	$\bar{1}$	0	1
$\bar{1}$	1	0	$\bar{1}$
0	0	0	0
1	$\bar{1}$	0	1

Task 7   1 $\bar{1}$ 1 $\bar{1}$ 0, 1 $\bar{1}$ 0 1 1 1, 1 $\bar{1}$ $\bar{1}$ 1 1 0 $\bar{1}$, 1 1 1 1 1.

5.5 'Choosing your weights' by Wendy Maull and Roger Porkess
Vol. 29, no. 1 (January 2000), pp 20-22

Introduction

You are given 3 weights and a pair of scales. This allows you to weigh any object whose mass is a whole number of grams, up to a certain limit. What are the sizes of the 3 weights and what is the limit?

This well-known puzzle gives considerable scope for classroom investigations that lead to a surprising range of interesting mathematical ideas. This article looks at some of the possibilities.

Notation

No doubt you know the answer to the given problem is that the weights are 1, 3 and 9 kg and with these you can weigh any object up to 13 kg. To weigh an object of mass 5 kg, for example, you would put the 9 kg weight on one side and those of 3 and 1 along with the object on the other side, as illustrated in Figure 1.

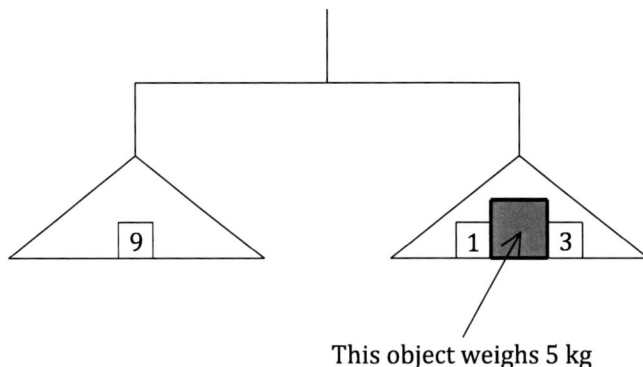

This object weighs 5 kg

Fig. 1

Students will want to satisfy themselves that they can indeed use these weights for any object up to 13 kg, but without the hassle of drawing scales each time. They need to devise a simple notation that will tell them which weights they are using and where, and the mass of the object. They might, for example, represent the diagram above by $9 - 1 - 3 = 5$. Many students will do this without thinking much about it. It is worth making the point that they have taken the very important step of translating a physical problem into mathematical notation.

What weight do you choose next?

Suppose now that you are allowed a fourth weight. What would you choose it to be?

The lightest object you have been unable to weigh so far is 14 kg. To weigh this you will need the new weight, W kg, on one side and the object, 14 kg, along with all the other weights, total 13 kg, on the other side.

$$W - 9 - 3 - 1 = 14$$
$$W = 14 + 13 = 27.$$

A similar line of argument shows the fifth weight to be 81 kg. So the sequence of weights is

$$1, 3, 9, 27, 81, \dots$$

This is a rather surprising result. You might well not have expected to have the powers of 3 crop up either from the problem itself or from the procedure for finding the next weight. How does it happen?

You can make the procedure for choosing further weights into a general formula. Call the total of all the weights you already have T kg. So the next object to be weighed must be $(T + 1)$ kg. On the one side you place your new weight W g and this equals the sum of the object and all the previous weights.

$$W - T = (T + 1)$$
$$W = 2T + 1.$$

So what is the connection between powers of 3 and the formula $W = 2T + 1$? The powers of 3 are the terms of the geometric sequence, $1, 3, 3^2, \dots$ The formula for the sum of the first n terms of a geometric sequence with first term 1 and common ratio 3 is given by

$$S_n = \frac{3^n - 1}{2}.$$

This can be rewritten as $3^n = 2S_n + 1$. This is the same as $W = 2T + 1$ since $S_n = T$ and $W = 3^n$.

Weights on only one side

So the algebra works out for powers of 3, but you might still feel there is a bit of a hole in the explanation. Why 3? Why not 2 or 4 or something else?

You can gain an insight into this question by looking at an alternative weighing arrangement in which all the weights must be on one side and the object on the other. (You can think of it as being hung from a hook rather than placed on a scale pan.)

In this situation the sequence of weights is the powers of 2: 1, 2, 4, 8, ... This result is useful in two quite different ways. On the one hand, it allows the work to be broadened out without becoming harder (you might well in fact present this as the starting problem and the previous case as an extension of it). On the other hand, it sets up a challenge. You have two sequences, one of powers of 2 and another of powers of 3; what weighing arrangement gives powers of 4?

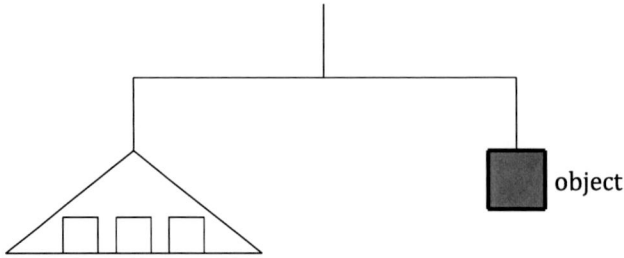

Fig. 2

Binary operations

In the situation we have just covered, there is just one weighing position (i.e. for the weights) and the weights go up in powers of 2. In the previous situation there are two weighing positions and the weights go up in powers of 3.

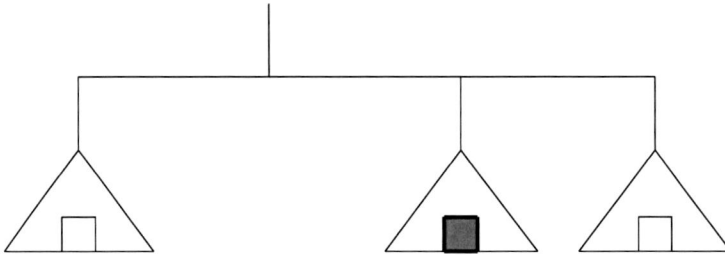

Fig. 3

A set-up with three weighing positions is shown in Figure 3. The new weighing position is on the object side of the sales twice as far out.

This is actually a rather complicated set-up but it allows an important step to be taken. This involves showing that it is equivalent to having the object on its own on one side (at distance 1), and three different weighing positions on the other, at distances 1, 2 and 3.

By adding in further weighing positions you can extend this idea to obtain sequences of powers of 5, 6, ..., indeed of any positive integer.

Mathematical justification

In this section, which is more theoretical, it seems preferable to describe the work in term of *numbers* rather than *weights*.

Suppose that there are n existing numbers, $W_1, W_2, W_3, \ldots W_n$ and that they cover all the possibilities up to their total, $T_n = W_1 + W_2 + W_3 + \cdots + W_n$. So the first T_n counting numbers are covered.

Suppose also that there are b weighing positions, all on one side. Each of these combines the next number W_{n+1} with each of the T_n combinations of the existing numbers or with zero. This gives $b \times (T_n + 1)$ new numbers.

So $T_{n+1} = b(T_n + 1) + T_n$

and since $W_{n+1} = T_n + 1$,

it follows that $W_{n+1} = (b + 1)W_n$.

Taken with $W_1 = 1$, this gives $W_n = (b + 1)^n$.

Efficiency

Another interesting feature of these situations is that the choices of weights are *efficient*. They allow you to weigh as many objects as possible with any number of given weights. In this section we restrict the discussion to the original situation with two scale pans and no extra weighing positions; however, the ideas are applicable to the other weighing arrangements.

Take the case of three weights. You can:

- Use one weight only to weigh 9, 3, 1   3
- Use two to weigh $9 \pm 3, 9 \pm 1, 3 \pm 1$ 6
- Use all three to weigh $9 \pm 3 \pm 1$ <u>4</u>

                                               13 objects

These are the only possible ways in which you can use the three weights, 13 in all, and these correspond to the 13 different objects, 1, 2, 3, ..., 13 kg. There can be no duplication.

To show that a set of weights is efficient, you need to establish that two conditions are met.

     A.   You can weigh every object, up to the maximum possible.
     B.   The number of weighable objects is the same as the number of possible arrangements of the weights. This is equivalent to saying there is no duplication; no object can be weighed using two different arrangements of the weights.

Thus the set 1, 3, 10 is not efficient because it fails condition A. It is not possible to weigh an object of 5 kg.

The set 1, 3, 8 is not efficient since there are 13 possible arrangements of three weights but $1 + 3 + 8 = 12$. In this case, 4 is duplicated: $3 + 1$ and $8 - 3 - 1$.

There are interesting teaching possibilities when proving that a set is efficient. You can use proof by exhaustion on a particular set, like 1, 3, 9, but this is only practical for small sets of known size. You cannot use exhaustion to prove that the general set 1, 3, 3^2, ..., 3^n is efficient. You can, however, give an inductive argument, thus:

The set 1, 3 is efficient: $1 = 1$, $3 - 1 = 2$, $3 = 3$, $3 + 1 = 4$. Adding 9 to the set gives the new set $9 \pm$ (1 to 4) and 9 itself, which is therefore also efficient. Such an argument can be generalized into a formal proof by induction. However, its classroom value is probably more in terms of helping students to develop concepts of proof, rather than in its detailed general application.

You can, rather messily, use a deductive argument to show that the greatest number of objects that you can weigh with n weights is $1 + 3 + 3^2 + \ldots + 3^{n-1}$.

No. of weights used	Number of ways of choosing weights	×	Number of ways of choosing their sides	=	Number of possible objects
1	$\binom{n}{1}$		$1 = 2^0$		$\binom{n}{1} \times 2^0$
2	$\binom{n}{2}$		2^1		$\binom{n}{2} \times 2^1$
3	$\binom{n}{3}$		2^2		$\binom{n}{3} \times 2^2$
...	...		...		...
$n-1$	$\binom{n}{n-1}$		2^{n-2}		$\binom{n}{n-1} \times 2^{n-2}$
n	$\binom{n}{n} = 1$		2^{n-1}		$\binom{n}{n} \times 2^{n-1}$

Thus when you have n weights, the greatest number of objects you can weigh is

$$\binom{n}{1} \times 2^0 + \binom{n}{2} \times 2^1 + \binom{n}{3} \times 2^2 + \cdots + \binom{n}{n-1} \times 2^{n-2} + \binom{n}{n} \times 2^{n-1}.$$

This is

$$\frac{1}{2}\left[\binom{n}{0} \times 2^0 + \binom{n}{1} \times 2^1 + \binom{n}{2} \times 2^2 + \cdots + \binom{n}{n} \times 2^n\right] - \frac{1}{2}\binom{n}{0}$$

$$= \frac{1}{2}(1+2)^n - \frac{1}{2}$$

$$= \frac{1}{2}(3^n - 1).$$

This expression is the same as $1 + 3 + 3^2 + \ldots + 3^{n-1}$. However, this is not sufficient to prove the process is efficient; for that you also need to show that there can be no duplication or that every object can be weighed.

Summary

We believe that this topic provides the basis for much interesting investigative work, and we have made some suggestions for paths along which students might be directed. However, investigative work can throw up challenging questions and we have also tried to point out some of the mathematical richness that may come to the surface.

Finally, a further extension to which we are not giving the answer.

> *In another situation the scale pans are small so that you can put no more than two weights on the one side and no more than one weight beside the object on the other side.*

> *(i) State the first four weights you would choose.*

> *(ii) Is this set efficient?*

5.6 'Domino tiling' by John Brundan
Vol. 23, no. 1 (January 1994), pp 23-24

I found the following investigation, from the new SMP 16-19 course, very interesting: "In how many different ways can you fill a rectangle measuring *m* units by *n* units with tiles shaped like dominoes which are 2 units long and 1 unit wide?" To illustrate what this means, there are exactly 11 ways of tiling a 3 by 4 rectangle in this way, as the diagram shows.

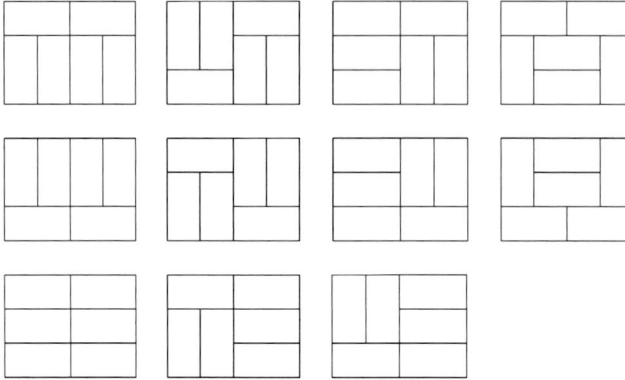

The investigation in this generality is a very hard problem, beyond sixth form level, but it does contain some interesting maths, and simple cases ("How many ways of tiling a 2 by 10 rectangle?") would definitely make a good investigation for younger pupils, with scope for work at all levels. I gave the problem of tiling rectangles 2 units wide to a top set year 9 class who had no problem spotting the number pattern. The task of drawing all possible tilings of a given rectangle, as here with a 3 by 4 rectangle, making sure none have been missed, is accessible to most pupils. It is impossible to tile for instance a 3 by 3 rectangle like this, or any other shape with *odd* area. Each domino has area two, so the total area covered by a tiling must be *even*.

A related puzzle is to tile a chessboard which has had two diagonally opposite corner squares cut off, using the same 2 by 1 dominoes. The chessboard certainly has an even number of squares, but still the tiling is impossible. The reason for this is just a different sort of parity. The corners cut away are the same colour, black in this case, so the truncated chessboard has more white squares than black. But, as each domino must cover exactly one white square and one black square, any tiling must cover the same number of white squares as black, and therefore it is impossible.

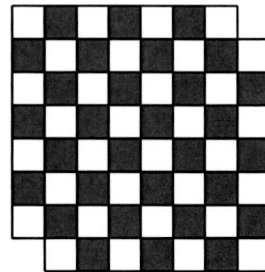

Taking the simplest case of rectangles 2 units wide and *n* units high, it is quite easy to work out the first few terms of the number pattern.

$n = 1$

$n = 2$

$n = 3$

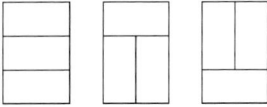

$n = 4$

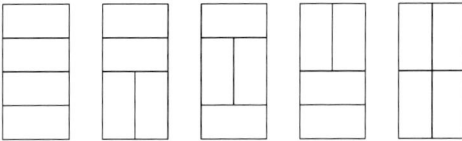

$n = 5$

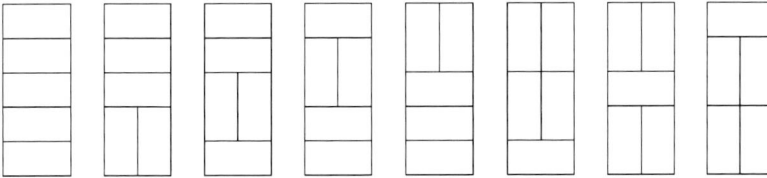

$n = 6$

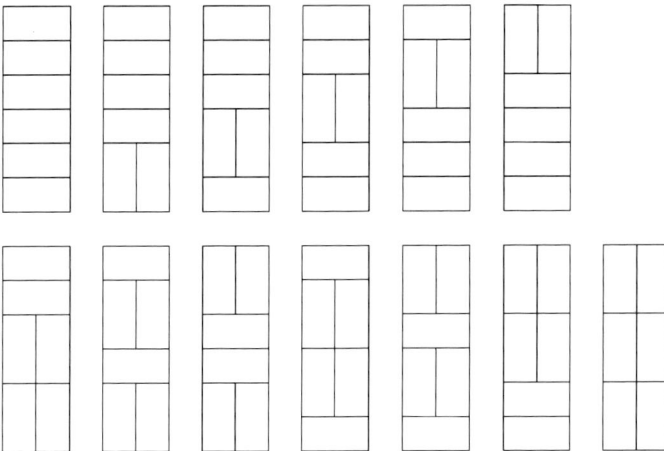

For $n = 1, 2, 3, 4, 5, 6$ the number of different tilings are 1, 2, 3, 5, 8, 13. This is the familiar Fibonacci sequence. The problem now is to explain why? To answer this, I used some recurrence relation notation. Let u_n be the number of ways of tiling the 2 by n rectangle. So, $u_1 = 1, u_2 = 2, u_3 = 3, u_4 = 5, u_5 = 8, ...$

Each tiling of a 2 by n rectangle must end either with a vertical domino, or with two horizontal dominoes. So, each of the tilings in the first case comes from a 2 by $(n-1)$ rectangle with a single vertical domino added on the edge; there are u_{n-1} of these altogether. Each of the tilings in the second case comes from a 2 by $(n-2)$ rectangle with a pair of horizontal dominoes added on the edge; there are u_{n-2} of these. Putting the two possibilities together gives the familiar recurrence relation for the Fibonacci sequence,

$$u_n = u_{n-2} + u_{n-1} \text{ with } u_1 = 1, u_2 = 2.$$

Intuitively, solutions are built up from shorter solutions by adding either one vertical or two horizontal dominoes to the end. So you can build from any of the previous two smaller solutions to get the next.

So far so good. Time to move on to 3 by n rectangles. We already know that only rectangles of even length will be possible this time, and that there are 3 ways of tiling the 3 by 2 rectangle, 11 ways of tiling the 3 by 4 rectangle. In fact, it is possible to set up a recurrence relation in almost the same way as for 2 by n rectangles. Whereas before, there were two possible endings, a single vertical domino or two horizontal tiles, now there are three endings to consider, corresponding to the three different ways of tiling a 2 by 3 rectangle.

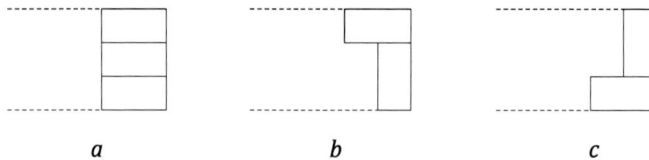

Let the number of tilings of a 3 by $2n$ rectangle with each of these endings be a_n, b_n and c_n respectively. So the total number of tilings of a 3 by $2n$ rectangle, u_n, is given by $u_n = a_n + b_n + c_n$.

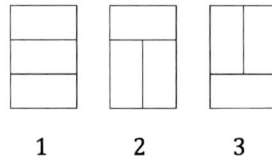

By symmetry, of course, $b_n = c_n$, so we will be able to simplify things later on. Now, just as before, I will look at how we can build up from a smaller tiling. This time, we will build up by adding one of three different blocks to each of the three possible endings. Each block is to be two units wide as there are no tilings of odd length. Block 1 can be added to any of the endings in precisely one way, and will generate a longer tiling with ending a), and this will account for all possible tilings of this length with this ending. So,

$$a_{n+1} = u_n = a_n + b_n + c_n = a_n + 2b_n.$$

Block 2 can be added in the same way to each of the endings to generate a longer tiling with ending b), but there is another possibility. If block 2 is added to ending b), the adjacent vertical dominoes that result can be flipped to produce a new sort of tiling:

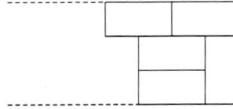

This has to be incorporated into the recurrence relation for b_n, so we get an extra term:

$$b_{n+1} = u_n + b_n = a_n + 3b_n.$$

Now we're there! The final recurrence relation is:

$$u_n = a_n + 2b_n, \text{ where } a_{n+1} = a_n + 2b_n, b_{n+1} = a_n + 3b_n \text{ and } a_1 = b_1 = 1.$$

To check, $a_2 = 3$, $b_2 = 4$ so $u_2 = 11$ gives the correct number of tilings of the 3 by 4 rectangle. For 4 by n rectangles, it gets harder, but still the recurrence method can be made to work. Of course, all of this is more complex than the investigation asked for, but I have been impressed by the amazing range of levels that this task can be approached – from year 9 to degree level.

5.7 Tony Orton on 'Fibonacci sideways'
Vol. 18, no. 3 (May 1989), p 27

Many years ago a mathematician known to us now as Fibonacci discovered a sequence of numbers, namely

$$1, 1, 2, 3, 5, 8, 13, 21, 34, 55, ...$$

which has fascinated explorers of mathematics ever since. Many of the properties and connections have been documented already. [Yet], ... there is certainly scope for each new generation of young mathematicians to investigate this sequence of numbers.

Nowadays it seems to be accepted that we may define other Fibonacci sequences, each constructed from a pair of starting numbers together with the rule that each subsequent term is formed by summing the previous two terms. Thus (2, 5) is sufficient to define

$$2, 5, 7, 12, 19, 31, 50, 81, ...$$

but (5, 2) defines a different sequence. Hence any number can be a term of a Fibonacci sequence, but the numbers in the sequence

$$1, 1, 2, 3, 5, 8, 13, 21, 34, 55, ...$$

are special, and therefore perhaps deserve to be known as 'the Fibonacci numbers'. Why might it be claimed that they are special?

The excellent investigation known as 'Fibonacci Backwards' opens up new lines of enquiry. It assumes that there can be many different Fibonacci sequences. It was explained in full by Gillian Hatch in 1982 (now available online, *Eds.*), but some readers might not have encountered it before, so here is a brief outline. If we are given a number in a Fibonacci sequence, say 47, what numbers precede it in the sequence so that the sequence is as long as it could possibly be? We might, for example, guess that the next number backwards is 27, and thus construct

$$47, 27, 20, 7, ...$$

When should this backwards sequence stop? Is there an alternative second number which would produce a longer sequence? Given any starting number, how do you work out what the next number backwards should be in order to produce the longest sequence?

We have used this particular investigation on many courses at Leeds and abroad for a number of years, and have observed that most investigators are so delighted when they think they have discovered a way of predicting the second number backwards that they imagine they have finished. Is a really good investigation ever finished? Two members of a recent course for Heads of Departments of Mathematics, having reached this stage fairly quickly, began to look sideways at their results, thus opening up new fields of enquiry. We therefore owe a debt of gratitude to Mike Sayles and John Ward. In the limited time available in session they succeeded in defining certain new questions and observed some properties unknown to us before. A computer program helped to speed up the investigations, but did not solve all problems. My own subsequent exploration has led to the formulation of a large number of questions which are worth a substantial sideways exploration from the forwards and backwards approaches which have constrained most students before.

1. All right, you think you have a rule for working out the second number in Fibonacci Backwards (if you haven't then that is the first thing to work on) but are you sure it always works? Have you double-checked that it always works? What questions does this checking raise in your mind?

2. The bigger the defined starting number in Fibonacci Backwards the more likely you are to be able to produce a long sequence, but the maximum length does seem to vary enormously for a given selection of consecutive natural numbers as starting numbers. Why is there such variation? What other questions emerge as you investigate?

3. As you work out the backwards sequences of maximum length for given starting numbers you eventually find the smallest number for which the maximum length is 10 (say). What is special about this number, the smallest number with maximum length 10? What is special about all numbers which are the first ones to have any particular maximum length?

4. Once a new maximum length first appears, say 10, it will appear again for some subsequent starting numbers. What is special about this set of starting numbers for which the maximum length is 10 (or any other number)?

5. Are there any starting numbers in Fibonacci Backwards which provide more than one answer to the problem of finding the longest sequence? What other questions does this lead to?

6. What is the relationship between starting number and maximum length of sequence in Fibonacci Backwards?

7. Did Fibonacci know even more than most modern books are able to tell us?

8. What other questions are worth asking?

Please don't expect me to provide answers because (a) this is an investigation and (b) I don't have very many as yet.

Reference

Hatch, G. (1982). *Jump to it!* City of Manchester College of Higher Education. The Editors have discovered that STEM Learning has made this booklet online via www.stem.org.uk/resources/elibrary/resource/30022/jump-it.

5.8 'Fraudulent dissection puzzles: A tour of the mathematics of bamboozlement' by John Sharp
Vol. 31, no. 4 (September 2002), pp 7-13; extract

Bamboozlement and hornswoggling are two wonderful sounding words which might have come out of Harry Potter, but they are terms applied to a type of dissection puzzle which is presented in the form of a paradox. They have been given to this type of puzzle by Greg Frederickson in his gathering of dissection puzzles, *Dissections Plane and Fancy* (Frederickson, 1997). Such puzzles have much to offer at all levels of mathematics with the added benefit of adding a bit of fun to mathematics. In this article, I want to take you on a tour of such puzzles, historically and mathematically, and to show that there are still uncharted waters waiting to be explored. The mathematics will take us through geometry and number sequences, with suggestions for investigations; and we will meet some mathematicians from our subject's history, known and not so well known, as well as an architect or two.

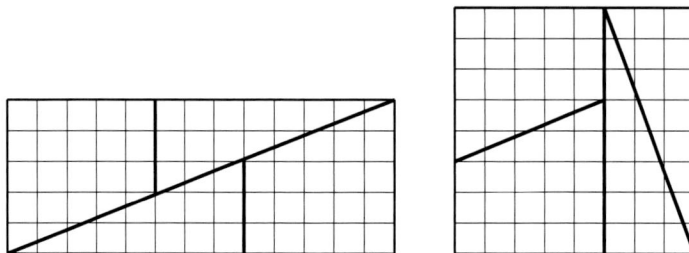

Fig. 1

The inspiration for the tour comes from the well-known dissection puzzle shown in Figure 1. This puzzle does not seem to have a name. It is usually described under a heading of 'a geometrical paradox'. I want to delve deeper into its mathematics and suggest how it might have been created or discovered. It offers potential for a variety of mathematics at many levels. It is normally given with a square of side 8 and a rectangle of sides 5 and 13. Anyone who has been exposed to the Fibonacci numbers will recognize them immediately and will probably know [Cassini's Identity], that for any two Fibonacci numbers, if $F_1 = F_2 = 1$,

$$F_n{}^2 = F_{n-1} \cdot F_{n+1} - (-1)^n \tag{1}$$

Since $F_6 = 8$, then the area of the square is less than the rectangle by one unit.

Lewis Carroll's notes on the puzzle

The puzzle seems to have appeared in this form about the middle of the 19th century. I have a theory as to how it arose which I will describe in a moment. It was a favourite of Lewis Carroll and among the papers unpublished in his lifetime are notes on a generalization of the problem, where he was the first to ask what other values can the sides of the square and rectangle have. The notes are incomplete and vague, but a possible reconstruction of his argument was published by Warren Weaver (Weaver, 1938). Using Carroll's method for seeing how the puzzle works, apart from providing some useful ideas for investigations, gives some mathematical clues to how the puzzle might have arisen.

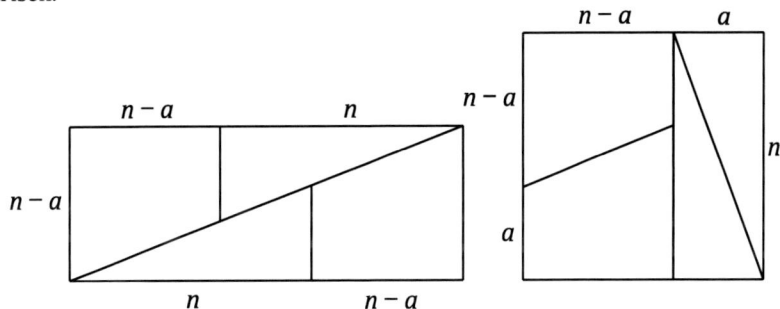

Fig. 2

Using the notation as in Figure 2, since the rectangle has an area one unit more than the square:

$$(2n - a)(n - a) - n^2 = 1 \qquad\qquad (2)$$

giving $\qquad\qquad a^2 - 3na + n^2 - 1 = 0 \qquad\qquad (3)$

Note that this equation is symmetrical in n and a and if we solve for a we get

$$a = \frac{3n \pm \sqrt{5n^2 + 4}}{2}.$$

Now we know we want a to be an integer, so there are lots of questions we can ask for an investigation.

Investigation 1

Find values of n for which a is an integer.

1. The expression above the line needs to be even to make a an integer. Does it make a difference if n is odd or even?
2. What values of n will result in an integer when the square root is taken?
3. Does the $\pm$ sign matter, or should only one of them be taken? In a class everyone could be given a value of n to calculate or, alternatively, you could set up a spreadsheet.

Table 1 shows integer results which answer these question.

n	Evaluation using + sign	Evaluation using − sign
1	3	0
3	8	1
8	21	3
21	55	8

Table 1

Using Lewis Carroll's method, we can thus see that there is an infinite number of solutions. You can then use these values to calculate the size of the rectangle. But note, because of the symmetry of Equation (3), that values of a take the same values as those for n, though of course not at the same time. Although the values form a series, they are not the Fibonacci series, but alternate terms of it. They are part of another recurrence sequence, however.

Suppose the rth value of this series is U_r. From the above table, note how, if a value of a is U_r, then the value of n is U_{r+1}. Now since a and n obey Equation (3) we can write

$$U_{r+1}^2 - 3U_{r+1}U_r + U_r^2 - 1 = 0 \tag{4}$$

$$U_r^2 - 3U_rU_{r-1} + U_{r-1}^2 - 1 = 0 \tag{5}$$

subtracting and factorizing

$$(U_{r-1} - U_{r+1})(U_{r-1} + U_{r+1} - 3U_r) = 0 \tag{6}$$

which means

$$U_{r+1} = 3U_r - U_{r-1} \tag{7}$$

Other versions of the puzzle

Figure 1 is a puzzle where the rectangle has an area one unit larger than the square. Equation (1) says that for the rectangle to be one unit more in area than the square, then the square must have the side of an even Fibonacci number, but it also says that if an odd one is used, then the rectangle has an area of one less than the area of the square. This means that Equation (3) becomes

$$a^2 - 3na + n^2 + 1 = 0 \tag{8}$$

If we apply the same reasoning as above, then we get another recurrence sequence in the equivalent of Table 1 (i.e. 1, 2, 5, 13, 34 ...) with the same relationship as Equation (7) but with different starting values. This can be used to extend the investigation.

There are other similar puzzles based on Fibonacci dissections. The pieces of Figure 1 can be rearranged as shown in Figure 3, so that using the results of Figure 1, it is possible to demonstrate that 63 = 64 = 65. This variation is sometimes attributed to the famous American puzzler Sam Loyd but, prolific though he was, he was a bamboozler too and often claimed other people's work as his own. The true inventor seems to have been Walter Dexter who published it in the *Boy's Own Paper* in 1901.

Figure 4 is a variation I found in one of my notebooks, but I do not know where it came from and is not mentioned in Frederickson (1997). It demonstrates that $13^2 = 8 \times 21$.

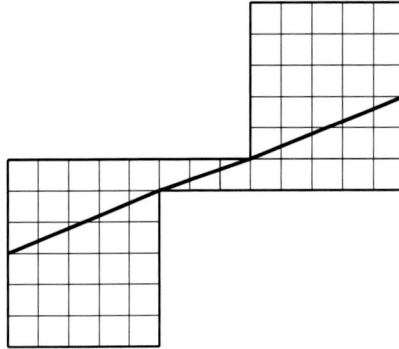

Fig. 3

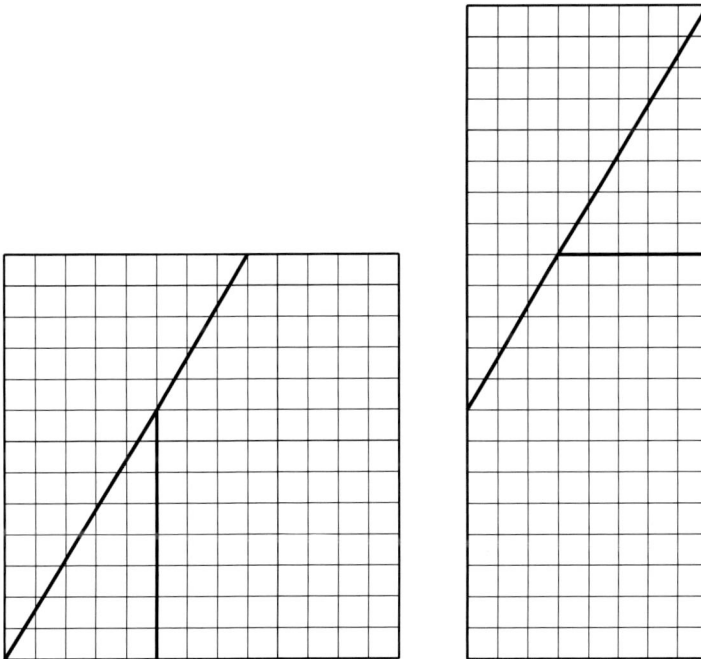

Fig. 4

Investigation 2

1. Can you find any more variations on the puzzle of Figure 1 using the same pieces?
2. What apparent area do you get in Figure 3 if you use the variation where the rectangle of Figure 1 has an area of one less than the area of the square?
3. Investigate the mathematics of Figure 4 using Lewis Carroll's methods.

203

How the puzzle might have been discovered

Rouse Ball (1892) gives a reference to the puzzle of Figure 1 which goes back to a German journal in 1868 but I have not been able to trace a copy. However, there is a similar puzzle which goes back to the 18th century and this may have an origin in a mistake of a 16th century architect, so let's start there and then see how the puzzle might have arisen from variations on this.

Sebastian Serlio and the table to door problem

Serlio (1475–1554) was an Italian architect who published a five volume work in 1551 which was the encyclopaedia of its day on architecture. It was translated into many languages and the English version of 1611 is still in print (Serlio, 1982). The first volume, *Libro Primo d'Architettura* is 'entreating of geometrie'. Folio 13 shows the diagram shown in Figure 5.

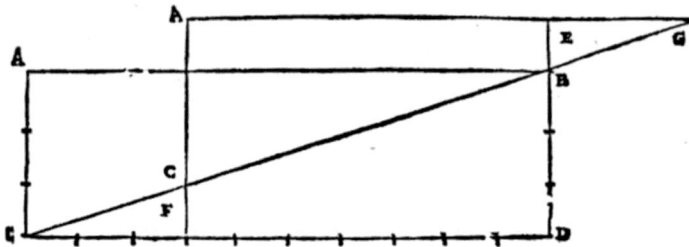

Fig. 5

The book is printed in a typeface that is hard to read, with the added complication of different spellings and somewhat quaint English. He describes the problem as follows.

> It may also fall out, that a man should find a table of ten foote long and three foote broade: with this Table a man would make a door of seven foote high and four foote wyde. Now to doe it, a man would saw the Table longwise in two parts and, setting them one under another, and so they would be but five foote high, and it should bee seven: and againe, if they would cut it three foote shorter, and so make it four foot broade, then the one side shall be too much pieced. Therefore he must do it in this sort: Take the Table of ten foote long and three foote broad, and marke it with A. B. C. D. then sawe it Diagonall wise, that is, from the corner C to B, with two equall parts, then draw one piece thereof three foote backwards towards the corner B, then the line A.E shall be foure foote broad and so shall the line E.D also hold foure foote broad: by this meanes you shall have your doore A.E.ED. seven foote long, and foure foote broade, and you shall yet have the three cornerd pieces marked E.B.G. and C.E and C left for some other use.

If we draw this and create the door from the table, then we have a set of diagrams like Figure 6.

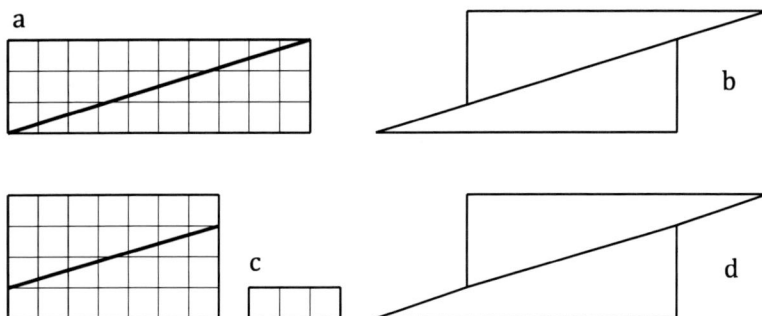

Fig. 6

The action would be to cut the table of Figure 6a and then slide it along so that you end up after cutting the triangular pieces off with two pieces like Figure 6c. Figure 6a has an area of 30 units, but Figure 6c has a combined area of 31, i.e. (28 + 3) units. Where has the extra area come from? Well, if you are exact in the way you get from 6a to 6c, then you see that there is a gap between the two triangles which is used to make the extra unit of length. So a true version of Figure 6c would have gaps along the diagonal also. I am pretty sure that Serlio did not know there was a problem. He was a practical man finding a solution to a need. It took another architect, Pietro Cataneo (c. 1510 – c. 1574), to point out the error and show that there was a paradox (Cataneo, 1567).

Investigation 3

Cataneo came up with an alternative dissection, using Serlio's method, of a 12 by 4 rectangle into a pair of rectangles of sides 9 by 5 and 3 by 1. Now, you will notice that there is no bamboozlement with these. Investigate what are the conditions for being able to get a perfect dissection or a paradox.

References

Cataneo, Pietro (1567) *L'Architettura di Pietro Cataneo Senese*, Venice.
Frederickson, Greg N. (1997) *Dissections Plane and Fancy*, CUP.
Rouse Ball, W. W. (1892) *Mathematical Recreations and Essays*, Dover.
Serlio, Sebastian (1611, 1982) *The Five Books of Architecture*, Dover.
Weaver, Warren (1938) 'Lewis Carroll and a Geometrical Paradox', *American Mathematical Monthly*, April, pp. 234-235.

Editors' note

What appears above represents perhaps half of what John Sharp wrote in his long and detailed article in *Mathematics in School*. So if you have enjoyed it to this point, you may wish to access the full version which takes the story on to more recent times.

6   *Miscellaneous Investigations*

6.0 Introduction

In this final Chapter we have gathered together some very good articles about investigations which do not really fit in with the themes of the previous five chapters. The first three articles have a hint of fractals in their investigations. This was a theme for investigations that was popular with mathematics masterclasses from the 1970s onwards. The article by Paul Scott, 'Taming the dragon', in Chapter 5 is another such example.

We thought that the contribution from David Miller, 'The case of the expanding spiral', is an excellent reminder of the sense of chaos and uncertainty that can happen in classes of students as they embark on an investigation which is fresh to a teacher. Unlike in the teaching of the traditional mathematics' topics, as teachers, we tend to know all the answers to questions that students might ask. For an investigation, we are never quite sure about the direction that students might go! This can be quite exciting for the classroom dynamic. Both teacher and student become 'learners' as the investigation progresses. David gives us this sense of following the unknown in his writing.

Sometimes an article in *Mathematics in School* suggests a new idea that has come out of an investigation. Sarah Perkins (now the Gresham Professor of Geometry, Sarah Hart) does this in her article proposing an extension to Euler's theorem for three-dimensional solids into four dimensions.

Problems which can lead to interesting investigations can often be developed from well know puzzles. In the first part of the 20th century Henry Dudeney and Sam Loyd published many puzzles and 'amusing tasks'. In his article Murray Macrae takes ten of these puzzles and develops them into investigations.

We complete this book of articles on the topic of investigations and their use in the classroom with an article by Roger Porkess, extending the idea of Pythagorean triples to non-right angles triangles. The conclusion to the article is a perfect opening paragraph to our book as well as a very appropriate way to finish:

> *Much concern has been expressed recently about students' lack of both algebra skills and understanding of proof. This piece of work addresses both within an essentially investigative setting; students who work through it stand to learn a great deal more than which particular triangles fit the given requirements.*

Written in 1998 – this is as true today as it was then!

6.1 'A spiral investigation' by A V Tourret
Vol. 12, no. 4 (September 1983), pp 2-3

A generous definition of a spiral is 'any object having a spiral form'. Using this definition, spirals may be very varied. They can be defined by the angle and the length and number of sides of the spiral. Look at the following varied examples of spiral diagrams.

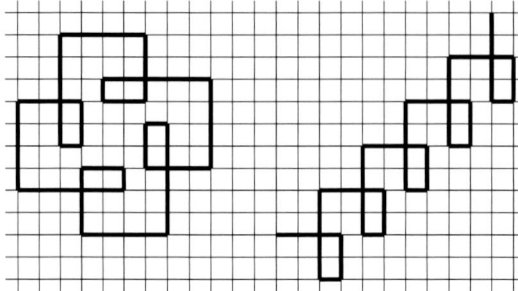

90° 1, 2, 3, 4, 5   **90° 1, 2, 3, 4**

60° 1, 2, 3, 4

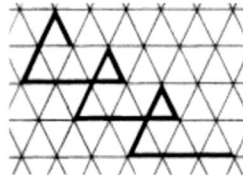

60° 1, 2, 3

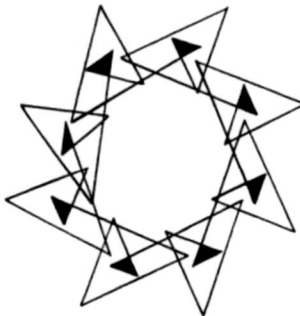

45° 1, 2, 3, 4, 5

It is possible to differentiate these spiral diagrams into two main categories: they are either continuous, such as 90° 1,2,3,4 and 60° 1,2,3 or they are enclosed and finite, such as 60° 1,2,3,4 and 45° 1,2,3,4,5. These examples indicated that there could be some interesting mathematics somewhere! Specialisation was carried out with the simplest finite spiral, i.e., 90°, with three sections of spiral a, b, c.

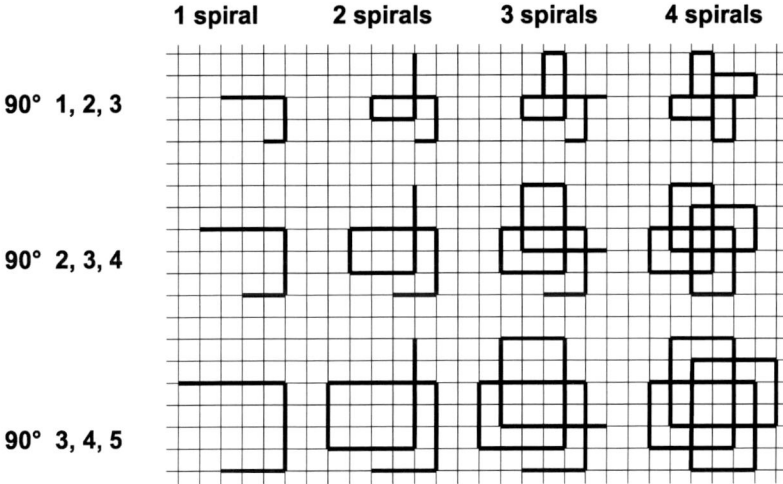

From these first specialisations, it became evident that the completed spiral diagram could be seen as being inscribed in a square. The dimensions of the circumscribed square could be obtained by adding the lengths of the two longer sides and subtracting the length of the shorter side. This hypothesis was tested with 90° 4,5,6 and found to be true for the completed spiral diagram was enclosed in a 7×7 square $(5 + 6 - 4)$.

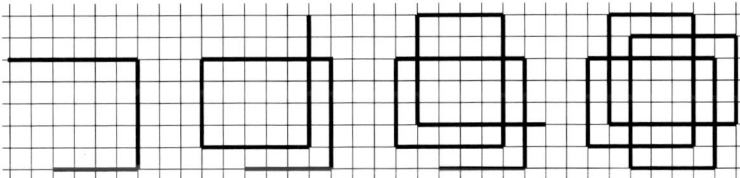

90° 4, 5, 6 (with 1, 2, 3 and 4 spirals)

Each of the spiral diagrams could be considered as four rectangles placed in a similar way with a circumscribed square with different areas of overlap.

The area of each of the rectangles within the spiral diagram could be obtained by multiplying the lengths of the two shorter dimensions, i.e., 4×5 in the 90° 4,5,6 spiral.

In all of the spiral diagrams investigated so far there were $4 \times (2 \times 1)$ square units which were not contained within the spiral diagram. Why was this so? To help solve this the following spiral was drawn: 90° 2,3,5.

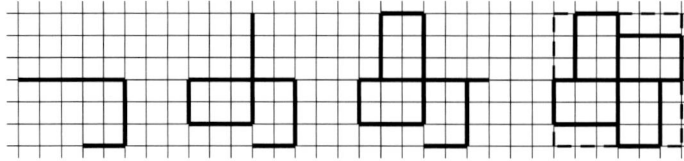

90° 2, 3, 5 (with 1, 2, 3 and 4 spirals)

The area of each of the four unused rectangles was 3 × 1 in this spiral. Looking back at the other spiral diagrams it became evident that the dimensions of the rectangles not contained within the spiral diagram could be obtained from the differences between the largest and the smallest number, and between the middle and smallest number. In the case of 90° 2,3,5 ...

$$5 - 2 = 3$$
$$3 - 2 = 1$$ giving 3 × 1 square units.

Another factor emerged from this spiral diagram – when the difference between the two larger dimensions was the same as the shortest dimension, there was no overlap. By now it was possible to derive for the 90° 1,3,4 that there should be no overlap. The circumscribed square would be 6^2, each rectangle would be 3 square units, and hence 24 squares would not be used, and these would be arranged in 4 × (3 × 2) square units.

The spiral diagram was as follows:

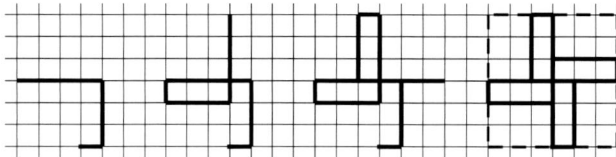

90° 1, 3, 4 (with 1, 2, 3 and 4 spirals)

So given the information 90° 2,4,5, the following hypothesis could now be made with confidence:

- Overlap would occur ($5 - 4 \neq 2$, smallest number 2).
- Circumscribed square would be 7^2 ($5 + 4 - 2 = 7$).
- Each rectangle within spiral diagram = 8 square units (4×2).
- Each rectangle not within spiral diagram
  = 3 × 2 square units $\big((5 - 2) \times (4 - 2)\big)$.

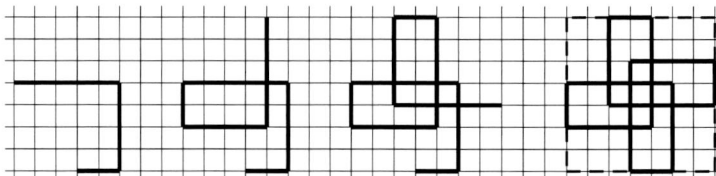

90° 2, 4, 5 (with 1, 2, 3 and 4 spirals)

It was now possible to generalise. Given $90°$ $n, n + a, n + a + b,$

- if $n = b$ there would be no overlap,
- area of circumscribed square $= (n + 2a + b)^2,$
- area of each rectangle within spiral diagram $= n(n + a) = n^2 + na,$
- area of each rectangle outside spiral diagram $= a(a + b) = a^2 + ab.$

This was only the beginning of the spiral investigation. Many questions were evoked from the early work relating to angles, further generalisation and proofs by induction.

6.2 Dave Miller on 'The case of the expanding spiral'
Vol. 17, no. 5 (November 1988), pp 42-43

It was a cold grey morning when the mag slapped onto the mat. The yellow of the cover demanded immediate attention. I ripped open the envelope and started on page 2 ('A spiral investigation' by A. V. Tourret). Facts and figures leapt from the page. I drew a 90 degree 1, 2, 3 spiral and let my mind wander. It didn't get very far but in this short time fate had primed my imagination. I carelessly tossed the mag aside. I had a journey to start and another day to survive.

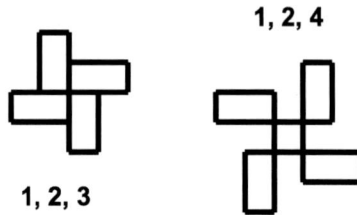

1, 2, 4

1, 2, 3

It was June when the ATM carelessly threw in their complications (in their usual way). It was a starting point, a worm of an idea. I squirmed, twisted and turned, then returned to my main source of excitement, marking examination papers. September found me needing an interesting case for 4A2. I found the mag lying where I'd left it a year earlier. Now was the time. They examined the Spiral evidence and came up with:

sum of largest pair take third gives minimum enclosing rectangle

Another case solved. I sat back and took a well-earned rest. Trouble comes in threes and sure enough Carl, Nicola and Rafiq were that trouble. "We want more than three numbers." They turned on the pressure, my knees went to jelly as they pounded at my eardrums. I gave in. "OK then, try it", I whined weakly. Peace at last. I needed the rest. Later they pummelled me with,

sixes repeat twice only, fives are like threes and fours never end.

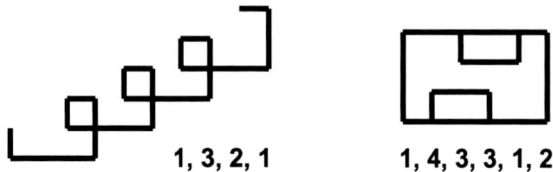

1, 3, 2, 1   **1, 4, 3, 3, 1, 2**

I'd have to get back my strength and face up to them. But later, not now.

Time passed and a new 4A2 came into being. This time I'd be prepared. "Investigate spirals", I said in a matter of fact way. There was little I didn't know. I'd remembered from last time! The smart crowd, Tracy, Linda and Mark came to me later with their solutions

*sevens are like threes, eights never end and the minimum square
enclosing a five is sum of largest three take the other two.*

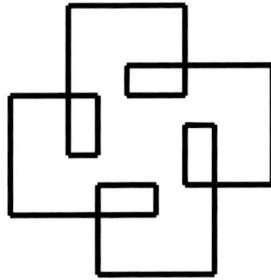

1, 2, 3, 4, 5

The dossier was still filling up. ... With 3B4 I started with the computer program. I loaded and fired "Draw that!" I emptied the whole chamber. Not a sound. The case was cracked. Drawings were rattled out in short staccato bursts; donuts, windmills, rects, pents and stars galore.

At the summer meet all the big shots were there. The Department sat back as I gave them the hard facts about the Spiral Case. "Phew, a close shave there", they relaxed over their straight decafs. "Thanks for the tip off. We'll be well prepared." Little did I know what was lurking in the rubber plants! I look back now at all the wrong theories I had and shudder. Spirals were to be executed in all public places. I had drawn both 3A1 and 3A5. Some deal!

I took 3A5 on first. Turning through 90 in the same direction each time created problems. A good job I had Sue and Liz as deputies! Try something else next time! Felt pens, the tables and chairs piled up, paper across the floor and footsteps made in red, blue and green felts. Teamwork was necessary. Success on the floor! Luckily the skill transferred after plenty of practice.

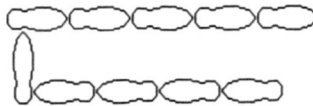

Footsteps 4, 1, 5: the start

All started well with 3A1: "Investigate all sorts of Spirals" was the task. They set feverishly to work. I let out clues 'smallest prison, repeat cycle, disappearance...'. Three lessons later the olde worlde order had been destroyed. Chaos began to reign.

Neil started the revolution. The smallest enclosing rectangle for a three spiral is sometimes given by the diagram shown here. The example he gave was for a 2, 3, 4 spiral; my counter blow (much later) was for a 3, 4, 5. This gives rise to a trainees' investigation, "when will the smallest enclosing rectangle lie along the grid lines?" Look out for rash deductions.

213

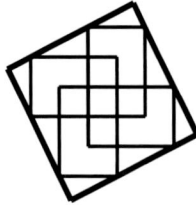

2, 3, 4 with minimum square   **3, 4, 5 with non-minimum square**

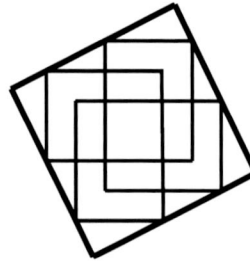

Richard then kicked me while I was down. "I couldn't find the grid line rule for five spirals." I checked his work. Catastrophe! It's a good job hanging had been stopped. My earlier juries must have only taken evidence from increasing spirals (e.g. 1, 3, 4, 6, 8 and 4, 5, 6, 8, 11 but not 6, 4, 7, 1, 5). So much for court injunctions.

I had to atone for these errors. A hair shirt and numerous evenings alone with all sorts of spirals gave me the solution for the general case for odd numbers only. The reader is now left to find the solution for herself.

Additional evidence has since been presented by SEG using negative numbers called 'Twists'. How about a common name. Was Ms Tourret first? *MT* 118 also feature Spirals and Worms.

Time passed and things changed. However I could not leave my favourite investigation alone and so I introduced it to my PGCE students. I might have known by now! Two different approaches appeared and one led to another theory being refuted. Vectors can be used (I say no more and leave you to join in the fun).

All spirals with number of lengths multiples of four always go on forever.

The initial counter was 1, 2, 3, 4, 4, 3, 2, 1 – thanks to Lesley and Martin. Using vectors supplies an infinite family of counters!

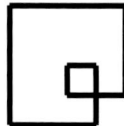

1, 2, 3, 4, 4, 3, 2, 1

So what now? I think I have general theories for isometric spirals, but I'm more cautious now! However, I know it was important to me at the start to believe that I knew all the results. Now I realise that the Spiral is forever expanding and that there is much more to be discovered.

References

ATM, *Points of Departure.*

Bloomfield, A (1987) 'Assessing investigations', *Mathematics Teaching* 118.

GCSE Mathematics - with centre based assessment. A Teachers' Guide. SEG.

Tourret, A V (1983) 'A spiral investigation', *Mathematics in School* 12, 4 (Sept.), pp 2-3.

6.3 R. Reed on 'The lemming simulation pattern'
Vol. 3, no. 6 (November 1974), pp 5-6

The pattern began (using isometric graph paper) with a single triangle and the first 'generation' was created by the addition of a triangle at each vertex. The second generation grew similarly. See Figures 1(i), (ii), (iii).

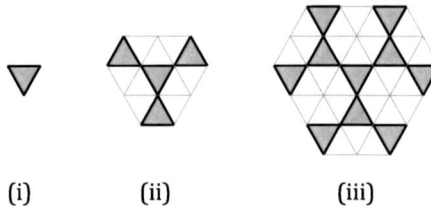

(i)   (ii)   (iii)

Figure 1

The third generation posed the first major problem: a choice had to be made; (a) should two 'parents' provide only one 'offspring' in a space? (b) could two parents provide two offspring in the same space? (c) should the space remain empty since any offspring would 'die' through overpopulation? Let us consider the first choice (a). The continuation of generations by the method already indicated produced a hexagonal pattern exhibiting three lines of symmetry. (See Figure 2.) The hexagonal border surrounding the nth generation offspring has sides of length n and $n + 1$ units. The number of offspring in the nth generation is $3n$, and the total number of triangles is P where

$$P = 1 + \sum_{r=1}^{n} 3r = 1 + \frac{3n(n + 1)}{2}.$$

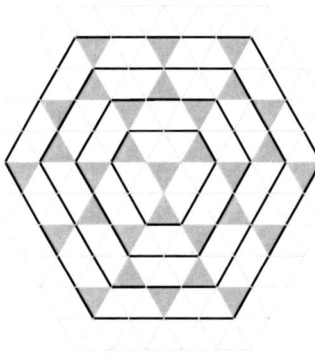

Figure 2

Now consider the second choice (b). By giving parents and offspring numerical values it is possible to show which offspring in a particular generation have more ancestors. For example, one parent, value 1, gives an offspring of value 1, whilst two parents, each of value 1, give an offspring of value 2; and, further, one parent value 2 and one parent value 3 give an offspring of value 5. By drawing the pattern with number values instead of colours, a startling result emerges!

216

Successive sides of the hexagon give rows of Pascal's Triangle. (Figure 3 shows the first few 'rings'. The original article showed more still. *Eds.*) The numbers are also the binomial coefficients.

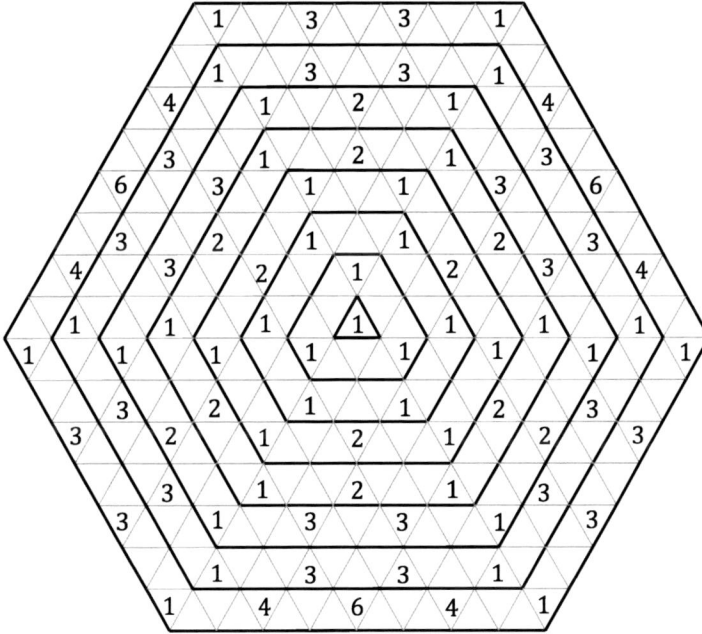

Figure 3

The number patterns obtained from this pattern are as follows:

Total Population:	1		4		10		22		40		70		112		178
Each Generation:		3		6		12		18		30		42		66		
First Difference:			3		6		6		12		12		24			

To construct an algebraic model for the number or triangles in each generation, n, consider the following.

n	Number of triangles
1	3
2	6
3	$3(1 + 2 + 1)$
4	$6(1 + 2 + 1) - 6 = 6(2^2 - 1)$
5	$3(1 + 2 + 1) + 3(1 + 3 + 3 + 1) - 6 = 3(2^2 - 1) + 3(2^3 - 1)$
6	$6(1 + 3 + 3 + 1) - 6 = 6(2^3 - 1)$
7	$3(1 + 3 + 3 + 1) + 3(1 + 4 + 6 + 4 + 1) - 6 = 3(2^3 - 1) + 3(2^4 - 1)$

which leads to these generalisations:

Generation $2n - 1$ has a population $3(2^{n-1} - 1) + 3(2^n - 1)$.

Generation $2n$ has a population $6(2^n - 1)$.

(Or, write the generalisations as $9 \cdot 2^{n-1} - 6$ and $6 \cdot 2^{n-1} - 6$, respectively. *Eds.*)

Hence the total populations P_{2n} and P_{2n+1} (for patterns with an even or an odd number of generations respectively) may be found as follows:

$$P_{2n} = 6 \sum_{r=1}^{n} (2^r - 1) + 3 \sum_{r=1}^{n} (2^{r-1} - 1) + 3 \sum_{r=1}^{n} (2^r - 1) + 1$$

$$= 9 \left[\frac{2(2^n - 1)}{2 - 1} - n \right] + 3 \left[\frac{2(2^{n-1} - 1)}{2 - 1} - n \right] + 1$$

$$= 21 \cdot 2^n - 12n - 20.$$

$$P_{2n+1} = P_{2n} + (2n + 1)\text{th generation} = 30 \cdot 2^n - 12n - 26.$$

Figure 4

Now consider the third choice (c), where parents are not allowed to 'breed' into the same space. In the tenth generation (see Figure 4) there is another choice to make from the following: (d) are parents only allowed to breed 'outwards'? or (e) are parents allowed to 'breed-back' into the previous generation, or even further back into the 'past'? What proves most fascinating about this particular pattern is that in the 8th, 16th and 32nd generations, only six 'live' cells remain and yet the 30th generation had been teeming with 90 live cells. Hence the allusion to lemmings in the title of this pattern; after a few generations the pattern destroys itself leaving only a few to carry on. The number of triangles in the first few generations is 3, 6, 6, 6, 12, 18, 6, 12, 6, 18, 18, 18, 30, 42, 246, ... [When graphed] the pattern can be split into 'spans', the nth span having 2^n generations Of these, the first 2^{n-1} are repeated as the first half of the next span; the second half having a direct relationship with the first half. It seems likely that an algebraic model can be set up to give the total population in any given span. [...] The pattern obtained with choice (d) which allows back-breeding is far denser than the previous pattern and gives the following sequence for total populations: 1, 4, 10, 16, 22, 34, 52, 64, ... and, as yet, no number pattern is forthcoming. Further work is necessary on the question of whether this last pattern is a 'true pattern' in that it would be possible to predict the population in any specified generation.

6.4 Sarah Perkins on 'Investigating four-dimensional figures'
Vol. 22, no. 23 (March 1993), p 46

Euler's theorem for three dimensional solids states that $V + F = E + 2$ where V is the number of vertices, F the number of faces and E the number of edges. When my sister, Mary, and I were thinking about this theorem, we wondered if perhaps there could be something similar in four dimensions, so we looked at a four dimensional hypercube.

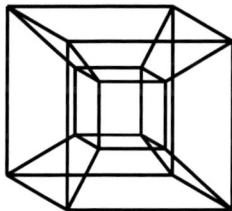

A four-dimensional hypercube

A four-dimensional hypercube is formed by joining the vertices of two cubes just as a cube is formed by joining the vertices of two squares. So obviously the hypercube has $2 \times 8 = 16$ vertices. A cube has 12 edges. In joining vertices 8 new edges are created, so the total number of edges if $(2 \times 12) + 8 = 32$ edges. From the diagram we could count 24 faces.

So, $V = 16$, $F = 24$ and $E = 32$. It appeared that $V + F = E + 8$.

We tried another four-dimensional shape, a 'hyperpyramid'. In three dimensions a pyramid consists of the vertices of a polygon joined to one point, so in four dimensions a pyramid is the vertices of a polyhedron joined to one point. We tried a cube-based pyramid.

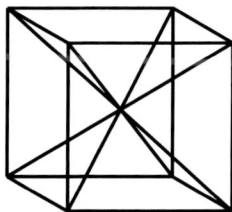

A cube-based pyramid

Here, $V = 9$, $F = 18$, $E = 20$. So in this case, $V + F = E + 7$.

So it wasn't just a matter of vertices, edges and faces. If in three dimensions you need to consider three things, then maybe in four dimensions, you need to consider four things. If vertices and edges define faces, and vertices, edges and faces define polyhedra, then vertices, edges, faces and polyhedra define four dimensional figures, so maybe the missing quantity is the number of three dimensional polyhedra (P) in the four dimensional figure. We drew up a table for various four dimensional figures of the quantities V, E, F and P.

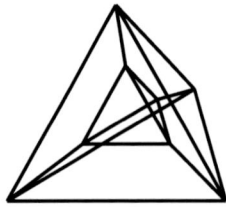

A four-dimensional tetrahedron

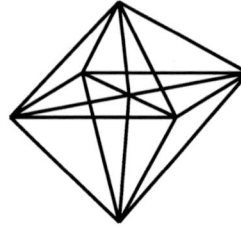

An octahedron-based pyramid

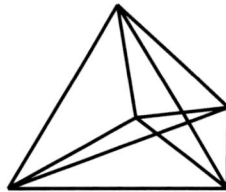

A tetrahedron-based pyramid

Shape	*V*	*F*	*E*	*P*
Four-dimensional cube	16	24	32	8
Cube-based pyramid	9	18	20	7
Tetrahedron-based pyramid	5	10	10	5
Four-dimensional tetrahedron	8	14	16	6
Octahedron-based pyramid	7	20	18	9

For these five figures at any rate, this formula is true:

$$V + F = E + P.$$

Although this formula worked for every shape we tried, we know that this doesn't constitute a proof! So perhaps readers could relate this to the three dimensional case by looking at projections into three dimensions, just as Euler's theorem can be proved by looking at projections into two dimensions.

6.5 W. Dagnall on 'Container routes'
Vol. 3, no. 3 (May 1974), pp 5-6

One of the standard puzzles discussed with students training for Junior Schools is the old favourite of the 3 pint and 5 pint jars that are used to measure out 4 pints. With one group of students I decided to use the information in Tom O'Beirne's *Puzzles and Paradoxes* (T. H. O'Beirne, OUP, 1965) and to approach the puzzles in a more devious way. The students were asked to adopt a dual role of responding to questions and situations, and of devising a suitable teaching method for children. The student's responses are included in brackets.

Using a large triangular grid, I drew an equilateral triangle of side 4 units. This was an arbitrary choice, and might have been 5 or 6 units for an introduction.

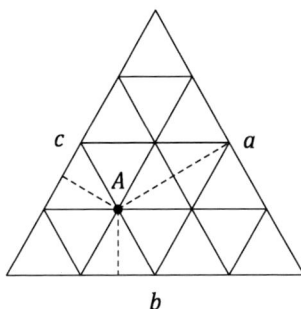

Figure 1

Students were given paper with a triangular grid and asked to draw the 4-unit triangle. "Choose any intersection and draw the altitudes to the sides of the main triangle."

"How many unit altitudes are there to each side?" (2, 1, 1)

"What is the sum of the altitudes?" (4)
(Would children know the word 'altitude'? We agreed to include in the lesson plan 'previous knowledge: altitude of triangle', to be checked or taught by the student.)

"Find the sum of the altitudes from other intersection points. What do you notice?" (By inspection, a constant sum of 4 vertices was quickly noticed, including the cases of the vertices of the large triangle.)

I did not mention points within the triangle other than intersection points, and this was not raised by the group. It could be given as an individual investigation if mentioned in class.

I marked the sides of the triangle *a*, *b* and *c* and asked for a record in brackets of the distances of a point from the sides. (There was no difficulty in recording, for example, *A* as (2,1,1) and vertices on the perimeter, such as (0,1,3) and (0,4,0). It was suggested that children should, be given practice of marking points given the co-ordinates. When I asked, "Why co-ordinates?", I was told that it was

221

obvious, "... like a graph". I suggested that the term 'number triple' might be used, but perhaps the word 'position' would do.)

I asked students to mark the points A (2,1,1) and B (0,1,3).

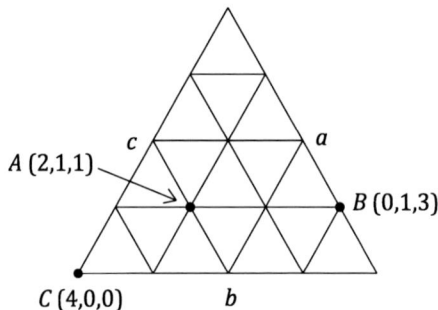

Figure 2

"What happens to the triple when moving from A to B?" (Some of the group noticed that the line AB is parallel to b, and that the distance from b is constant. Asked for a generalisation, I was told that movement between any two points would be parallel to one side and the distance from that side in the triple would be unchanged.)

Now came the time to study the path of an imaginary billiard ball. "Imagine a billiard ball at a point on the perimeter. Which point?" (3,1,0) was chosen. "Hit the ball parallel to b and record its route as it bounces off the cushions that bound the triangle." (A record was made: (3,1,0)→(0,1,3) → (1,0,3) → (1,3,0) → (0,3,1) → (3,0,1) → (3,1,0), and then around again. When asked for any comments about the pattern, the group came up with a sum of 4, and always a 0 in the triple, with the 0 in a different position each time.)

"What happens if the ball starts from a vertex?" (One student chose C (4,0,0) and built up a chain of (4,0,0) → (0,0,4) → (0,4,0) → (4,0,0) and so on.) This line of approach could have been continued by examining triangles with sides of a different number of units, but did not seem very productive in leading towards the initial puzzle. If we start at a vertex of the large triangle then the path goes around the perimeter, while if we start at a non-vertex point then the path never reaches a vertex. This is not to suggest that the routes using a full triangle are of little interest. Angles of incidence and reflection, and length of routes, (parallel to each line and total length), are worth consideration.

"Let's cut off one corner of the triangle, leaving a trapezoid with distance from c not greater than 2 units. Choose a point on the perimeter. What is its position?" (One point chosen was A (0,3,1).) "Trace a route from A and write down the positions where the route changes direction." (The first route chosen was (0,3,1) → (3,0,1) → (3,1,0) → (1,1,2) → (1,3,0) → (0,3,1). The analogy with a billiard ball route is suitable here. The next suggestion was a route around the perimeter: $(0,3,1) \rightarrow (0,4,0) \rightarrow (4,0,0) \rightarrow (2,0,2) \rightarrow (0,2,2) \rightarrow (0,3,1)$. No other suggestions for routes were made.)

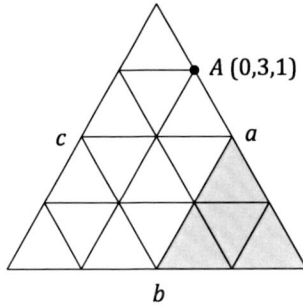

Figure 3

I suggested that at a corner of the trapezoid the route could be continued around the perimeter or inwards from the vertex, or the reverse of this.

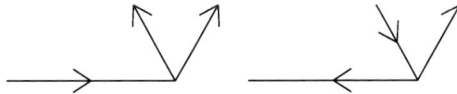

Figure 4

(There was a strong protest that this was not a billiard ball path, and the group appeared to regard my new rule as arbitrary and unnecessary. Perhaps I should have thought of a different approach to the puzzle that did not involve billiard ball routes. Or have I missed a simple way of adapting the analogy? Anyway, new routes were suggested using the new rule, for example, $(0,3,1) \rightarrow (0,2,2) \rightarrow (2,2,0) \rightarrow (2,0,2) \rightarrow (4,0,0)$, and on reaching a vertex of the triangle I created a pocket into which the ball dropped.)

"Let's try this with a larger triangle, and this time cut off two vertices. Work on your own, or with a group." One group chose a triangle of 5 unit side and cut off two vertices so that the distance from b was not greater than 4 and the distance from c not greater than 3.

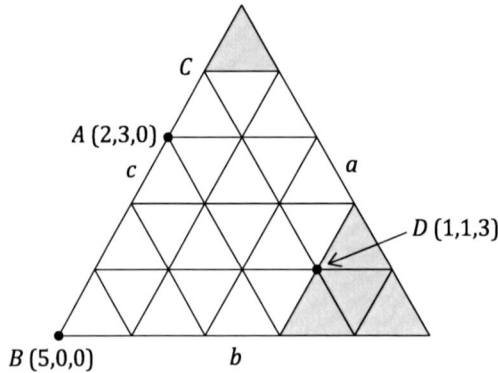

Figure 5

A point A (2,3,0) was chosen as the starting point, and I asked if the group could find and record a route from (2,3,0) to the vertex (5,0,0). There was another difficulty here. I wanted to lead to the idea of a route which when reversed would start at B (5,0,0) and end at A (2,3,0), with A at the end of a line of the route (so the route cannot end at A via a last line BA or CA). It was suggested that we say "full line route" and explain to the children what is meant by a 'full line'. Several full line routes from A to B were found and recorded, and I then asked for the shortest route, with mutterings of "It's probably Pascal's Triangle again". The shortest route found was (2,3,0) → (2,0,3) → (5,0,0).

"Now choose a point on the trapezoid that is not on the perimeter of the triangle. Find a shortest route from this point to B (5,0,0)." Following on with the same group, a point D (1,1,3) was chosen, with a shortest route of (1,1,3) → (1,4,0) → (5,0,0).

At this point I decided to lead towards the idea of capacities of containers. I asked the group to imagine that the maximum distances from the lines a, b and c represented the capacities in litres of three containers, i.e. a 5 litres, b 4 litres and c 3 litres. I said that we could pour from one container to another but that there were no graduations on the containers. When pouring from one container to another, one must start empty or full or finish empty or full. We establish that the point (5,0,0) represents the 5 litre jar full and the two other jars empty, and the meaning of a move (5,0,0) → (1,4,0) as filling jar b from jar a.

"How can we end with the jars holding 1 litre, 1 litre and 3 litres?" (Make the moves (5,0,0) → (1,4,0) → (1,1,3).) I went back to the diagram of the 4 unit side triangle, and asked for the maximum capacity of the containers.

224

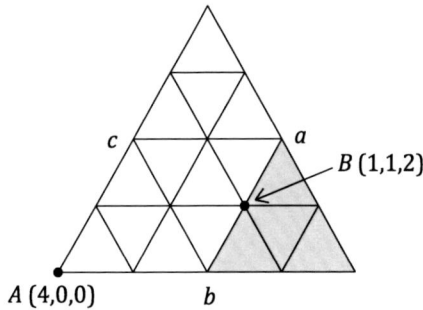

Figure 6

(A quick response was *a* 4 litres, *b* 4 litres and *c* 2 litres.) "Starting with the *a* jar full and the two others empty, how can we end with 1 litre, 1 litre and 2 litres in the jars?" (A backtrack from (1,1,2) led to (1,1,2) again without passing through (4,0,0). The group was ready to concede that the pouring cannot be done.)

Had time permitted, we might have constructed triangles, shaded parts of the triangles and devised container puzzles based on these triangles. Coffee break loomed, and I quickly gave the puzzle of the 8-litre, 5-litre and 3-litre jars. Start with the 8-litre jar full, and finish with 4 litres in each of two jars. When I returned, stronger hearts than mine had solved the puzzle.

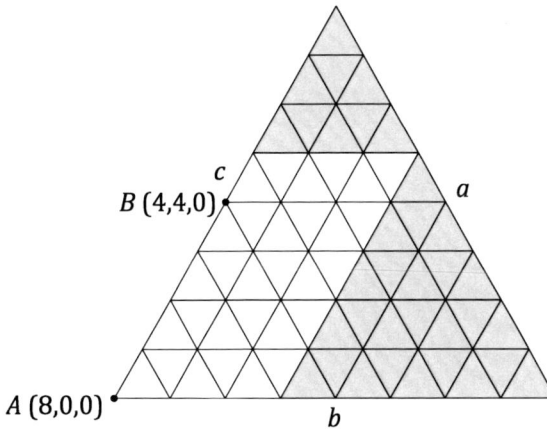

Figure 7

The routes found were:

a) (4,4,0) → (4,1,3) → (7,1,0) → (7,0,1) → (2,5,1) → (2,3,3) → (5,3,0) → (5,0,3) → (8,0,0): 8 moves.

b) (4,4,0) → (1,4,3) → (1,5,2) → (6,0,2) → (6,2,0) → (3,2,3) → (3,5,0) → (8,0,0): 7 moves.

The reversals of these routes give the order of pourings to end with 4 litres in two of the jars.

225

(There was the observation that 3, 5 and 8 are consecutive terms of the Fibonacci Series and I was asked if there was a relation. I do not know of one, but as the next suitable Fibonacci triples are (13,21,34) and (55,89,144) this line of enquiry does not seem very profitable for the classroom. Have I missed an obvious relationship?)

Two of the students decided to try the idea on their weekly morning visit to a 4th year junior class. They reported a fair measure of success and considerable interest in devising and solving container puzzles based on triangles with up to 6-unit sides. The 3, 5, 8 puzzle was not mentioned to the children. O'Beirne's chapter on containers offers much more than the limited section that we developed. The section appeared worthwhile in giving a general method of solution of container puzzles, and in offering another method of fixing position by a triangular frame of reference.

6.6 Murray Macrae suggests '10 investigations based on the puzzles of H E Dudeney and Sam Loyd'
Vol. 21, no. 3 (May 1992), pp 36-39

In two compendiums published during the First World War, H. E. Dudeney (1917) and Sam Loyd (1914) offered a huge collection of puzzles and amusements, virtually all with a mathematical flavour. From the UK and USA respectively, Dudeney and Loyd were the forerunners of the authors of present day puzzle corners and brain teasers. Much of their work was re-published by Dover in the 50's and 60's under the editorship of Martin Gardner (1958, 1959, 1960). Their puzzles continue to re-appear in the pages of popular books, a recent collection edited by Carter and Russell (1990) being one of many. Dudeney and Loyd mostly presented their puzzles as outright challenges to their readers. This article suggests ways in which some of those challenges can be extended into investigations. The extended versions may be suitable either for recreational purposes or for schools or both. (As creators of problems for recreation, Dudeney and Loyd may well have preferred them not to be institutionalised!) The two compilers presented their puzzles in a style indicative of the times in which they wrote. In the following, the original wording has been retained where appropriate and is shown within quotes. Original diagrams have been copied and references are given to the puzzle numbers as they first appeared.

1. The four lions [Dudeney, 1917, No 296]

"The puzzle is to find in how many different ways the four lions may be placed so that there shall never be more than one lion in any row or column."

Investigations

(a) What does 'different' mean in this context?
(b) What if there were just 2 lions? ... or 3 lions?
(c) What if the four lions were placed in a 4×5 rectangular grid?, ... or a 4×6? ...
(d) Investigate for 2×2, 3×3, 5×5 square grids; also for rectangular grids.
(e) Generalise for m lions on an $n \times n$ grid $(m \leq n)$.

2. Count the votes [Loyd, 1914, No 90; Gardner, 1959]

In an election "5219 votes were cast for four candidates. The victor exceeded his opponents by 22, 30 and 73 votes, yet not one of them knew how to figure the exact number of votes received by each. "Can you give a simple rule for obtaining the desired information?"

Extension

Extend the rule for a typical modern by-election in which there are *n* candidates of parties such as the Monster Raving Looney Party, the Corrective Party and the Fancy Dress Party.

3. The digital century [Dudeney, 1917, No 94]

$$1 \; 2 \; 3 \; 4 \; 5 \; 6 \; 7 \; 8 \; 9 = 100$$

"It is required to place arithmetical signs between the nine figures so that they shall equal 100. Of course you must not alter the present numerical arrangement of the figures. Can you find a correct solution that employs

(1) the fewest possible signs

(2) the fewest possible strokes or dots of the pen?"

Investigations

Dudeney gave many solutions. The most economical and therefore the most elegant was his "singularly simple", $123 - 45 - 67 + 89 = 100$, which he doubted would ever be beaten. The problem can be extended by considering shorter sequences of digits:

(a) 1 2 3 = *n*. Place arithmetical signs between the digits so that

(i) *n* = 0, 1, 2, 3, 4, 5, 6, 7, ..., (ii) *n* is a maximum, (iii) *n* is a minimum. Then investigate for other runs of consecutive digits.

(b) 1 2 3 ... *k* = 100. Find the least value of *k* such that with arithmetic signs between the digits the outcome is 100. Find values of *k* for which this task is impossible.

4. Archery puzzle [Loyd, 1914, No 92; Gardner, 1960]

"How many arrows does it take to score exactly 100 on this target?"

Investigations

(1) An archer shoots three arrows. One lands in the outer two rings, one lands in the inner two rings and the last lands in the remaining two rings. Find all the possible scores. Investigate for other combinations of rings and arrows.

(2) Design a 6-ring target such that
   a) each ring contains a 2-digit score,
   b) each score is different
   c) each score is 30 or under
   d) it would be impossible to score a total of 100, no matter how many arrows landed on the target.

5. The five dominoes [Dudeney, 1917, No 379]

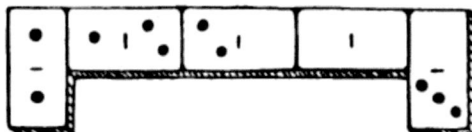

"It will be seen that the five dominoes are so arranged in proper sequence, that the total number of pips on the two end dominoes is five and the sum of the pips on the three dominoes in the middle is also five. There are just three other arrangements giving five for the additions". They are

(1-0)	(0-0)	(0-2)	(2-1)	(1-3)
(4-0)	(0-0)	(0-2)	(2-1)	(1-0)
(2-0)	(0-0)	(0-1)	(1-3)	(3-0)

"Now, how many similar arrangements are there of five dominoes that shall give six instead of five in the two additions?"

Investigations

(a) Investigate for different totals, e.g. 7, 8, 9, ...
(b) Investigate for shorter and longer chains of dominoes.

6. How much wire will it take? [Loyd, 1914, No 145; Gardner, 1960]

"I will give a little puzzle that I was called upon to tackle the other day. I found an electrician, who had invented some kind of switchboard, trying to work out the most economical method of stringing a fine copper wire through all the contact points on his board. The board was an elaborate affair, consisting of several hundred points, but since 64 is sufficient to illustrate our problem, only an 8 by 8 section of the board is shown.

"The problem is to find the shortest length of wire that will go from point B to the center of the little square marked A. The wire must touch the centers of all

64 little squares. Each square is one inch wide, and they are spaced so their centers are three inches apart. Each time a wire turns a corner, it is necessary to wind it around a corner of a square, an operation that uses two inches of wire. No diagonal connections are permitted.

"Assuming that two inches of wire are used in going from B to the center of the nearest little square, can you determine the shortest length of wire required to go from B to A?"

Investigations

(a) With B and A in the relative positions shown, solve the problem for boards of size from 3 by 3 to 10 by 10. Is it possible to find solutions for every configuration?

(b) What if B is in the top right-hand position and A is, as shown, in the bottom left-hand position?

(c) Investigate wiring diagrams for various positions of A and B.

(d) The puzzle (and the investigations) can be simplified without loss by ignoring the intricate windings at the terminals. For example, the following diagram shows two ways of connecting A to B on a 4 × 4 grid.

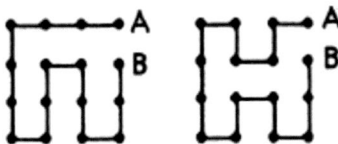

Are there any other ways? How many? What if A and B were in relatively different positions? ... or on a 5 × 5 grid?

7. A counter puzzle [Dudeney, 1917, No 326]

"Copy the simple diagram, enlarged, on a sheet of paper; then place two white counters on the points 1 and 2, and two red counters on 9 and 10.

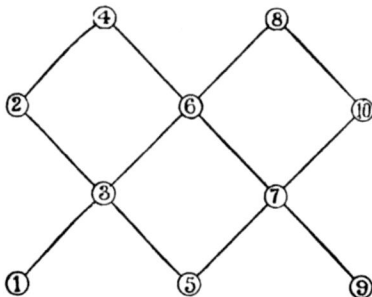

"The puzzle is to make the red and white change places. You may move the counters one at a time in any order you like, along the lines from point to point, with the only restriction that a red and a white counter may never stand at once on the same straight line. Thus the first move can only be from 1 or 2 to 3, or from 9 or 10 to 7"

Investigations

(a) What if the restriction is removed?
(b) What if reds and whites may be in a straight line but red and white counters
    must be moved alternately?
(c) Investigate games like this on similar boards.

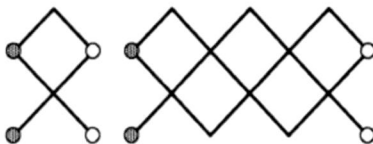

8. How can Rip Van Winkle win the game?
   [Loyd, 1914, No 6; Gardner, 1960]

"The old Dutch game of Kugelspiel used to be played with thirteen pins placed
in a row. [Either one pin or two adjacent] pins could be knocked down by any
single shot. Players bowled alternatively, one ball at a time, and the point of the
game was to see who could knock down the last pin.

"The little Man of the Mountain, with whom Rip Van Winkle is playing this game,
has just rolled a ball and knocked down pin No. 2. Rip has a choice of 22 different
plays: any one of the 12 single pins or any one of the ten open spots that will
bring down two adjacent pins. What is Rip's best shot to win the game? It is
assumed that both players can hit any pin or pair of pins they wish, and that
there is the best possible play on both sides."

Investigations

(a) Investigate this game for 1, 2, 3, ... *n* pins.
(b) What is the best strategy?
(c) Is there an advantage in starting first?

9. Five jealous husbands [Dudeney, 1917, No 375]

"During certain local floods five married couples found themselves surrounded by water, and had to escape from their unpleasant position in a boat that would only hold three persons at a time. Every husband was so jealous that he would not allow his wife to be in the boat or on either bank with another man (or with other men) unless he was himself present. Show the quickest way of getting these five men and their wives across safely.

"Call the men A, B, C, D, E and their respective wives *a,b,c,d,e*. To go over and return counts as two crossings."

Investigations

(a) Investigate for 2 couples, 3 couples and so on up to *n* couples.
(b) What if there were 5 couples, no jealous husbands but the boat could hold no more than two women and one man at any one time? (Assume that any one man weighs more than any one of the women.)
(c) Would it be possible to cross the river if the husbands were jealous and if the above weight restriction also applied?

10. Which is the best play and how many boxes will it win?
[Loyd, 1914; No 91, Gardner, 1959]

The game of 'dots and lines' is so familiar that it hardly needs to be explained. It is a contest between two or more players, to create the most square cells or boxes by joining dots orthogonally. Players take turn to draw a line. If a player 'scores' a box, he or she plays again. Loyd's puzzle shows a stage in a game:

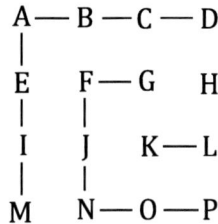

$$A — B — C — D$$
$$| $$
$$E \quad F — G \quad H$$
$$| \quad |$$
$$I \quad J \quad K — L$$
$$| \quad |$$
$$M \quad N — O — P$$

"The little maiden sitting down has to play. What play would you advise, and how many boxes will it win against the best possible play of the second player?"

Investigations

(a) Analyse strategies for play on the following grids. Is there an advantage in going first?

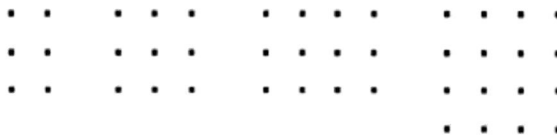

(b) Extend your analysis to grids which produce

(i) triangular cells:

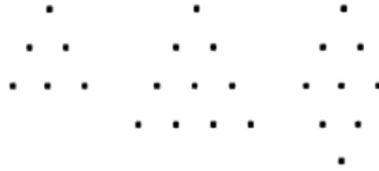

(ii) hexagonal cells.

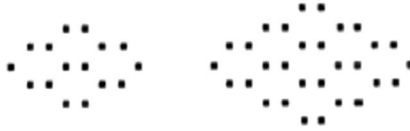

A re-reading of Dudeney and Loyd will produce many rewards for those seeking starting points for investigations. Their old bottles contain much new wine.

References

Carter, Philip and Russell, Ken (1990). *Classic Puzzles* Sphere Books, Macdonald.

Dudeney, H. E. (1917). *Amusements in Mathematics*, Thomas Nelson & Sons Limited; re-published in 1958 by Dover, New York.

Gardner, Martin (ed) (1959). *Mathematical Puzzles of Sam Loyd, 1*, Dover, New York.

Gardner, Martin (ed) (1960). *Mathematical Puzzles of Sam Loyd, 2*, Dover, New York.

Loyd, Sam (1914). *Cyclopedia of Puzzles*, 283 of which were re-published in Gardner (1959, 1960), after selection and editing.

6.7 'Sums of consecutive integers' by John Parker

Vol. 27, no. 2 (March 1998), pp 8-11

The problem of which numbers we can make by adding together two or more consecutive (positive) integers, is one of a number of mathematical problems which lend themselves to investigation at several different levels, and my purpose here is to pose, and suggest answers for, various questions which may engage and stretch the mathematical powers of students of any age and ability up to year eleven and beyond. At a very simple level the problem serves the modest purpose of introducing pupils to the ideas of 'consecutive' and 'integer', and of giving practice in adding. Deeper investigation will see us looking closely at the overall pattern of results with a view to formulating some general rules: later we shall consider the algebra of the problem and suggest proofs of why some numbers can and others cannot be made in this way.

A Starter Worksheet

If we add together two or more consecutive positive integers, their sum will be a positive integer:

$$2 + 3 = 5; \quad 3 + 4 + 5 = 12; \quad 5 + 6 + 7 + 8 = 26.$$

Can you find any integers which cannot be written as the sum of consecutive positive integers? Can you find any which can be so written in more than one way? Investigate.

One way to begin is to add integers in this way and tabulate the results. Full results as far as 30 are:

1		18	$5 + 6 + 7$
2			$3 + 4 + 5 + 6$
3	$1 + 2$	19	$9 + 10$
4		20	$2 + 3 + 4 + 5 + 6$
5	$2 + 3$	21	$10 + 11$
6	$1 + 2 + 3$		$6 + 7 + 8$
7	$3 + 4$		$1 + 2 + 3 + 4 + 5 + 6$
8		22	$4 + 5 + 6 + 7$
9	$4 + 5$	23	$11 + 12$
	$2 + 3 + 4$	24	$7 + 8 + 9$
10	$1 + 2 + 3 + 4$	25	$12 + 13$
11	$5 + 6$		$3 + 4 + 5 + 6 + 7$
12	$3 + 4 + 5$	26	$5 + 6 + 7 + 8$
13	$6 + 7$	27	$13 + 14$
14	$2 + 3 + 4 + 5$		$8 + 9 + 10$
15	$7 + 8$		$2 + 3 + 4 + 5 + 6 + 7$
	$4 + 5 + 6$	28	$1 + 2 + 3 + 4 + 5 + 6 + 7$
	$1 + 2 + 3 + 4 + 5$	29	$14 + 15$
16		30	$9 + 10 + 11$
17	$8 + 9$		$4 + 5 + 6 + 7 + 8$
			$6 + 7 + 8 + 9$

First conclusions

These early results suggest that:

1. No power of 2, including $2^0 = 1$, can apparently be obtained in this way.

2. All odd numbers from 3 onwards can be written as the sum of *two* consecutive integers.

3. The triangular numbers, T_n, appear clearly in the table, each being the sum of consecutive integers starting with 1, e.g. $T_4 = 10 = 1 + 2 + 3 + 4$.

4. All multiples of 3 from 6 onwards can be written as the sum of 3 consecutive integers; the middle integer is one-third of the number, e.g,. $18 = 5 + 6 + 7$, and 6 is one-third of 18.

5. All multiples of 5 from 15 onwards can be written as the sum of 5 consecutive integers; the middle integer is one-fifth of the number.

6. All even numbers from 10 onwards which give a remainder of 2 when divided by 4 (that is, which are of the form $n \equiv 2$, modulo 4) can be written as the sum of 4 consecutive integers. [Here we introduce the idea of congruence to a modulus. The set of numbers 10, 14, 18, 22, 26, 30,... all give a remainder of 2 when divided by 4, and we say they are all congruent to 2 when the modulus (the divisor) is 4, or to 2 (mod 4) for short.]

7. Likewise all multiples of three from 21 onwards which give a remainder of 3 when divided by 6, that is, which are of the form $n \equiv 3$ (mod 6), can be written as the sum of 6 consecutive integers. At a slightly deeper level, we can generalise some of these results.

8. All multiples of any odd number $(2n - 1)$ from $n(2n - 1)$ onwards can be written as the sum of $(2n - 1)$ consecutive integers; the middle integer is one-$(2n - 1)$th of the number.

9. All multiples of n from $n(2n + 1)$ onwards which give a remainder of n when divided by $2n$, that is, which are of the form $m \equiv n$ (mod $2n$), can be written as the sum of $2n$ consecutive integers.

A question

The smallest number that can be written as the sum of consecutive integers in two different ways is 9; in three different ways is 15; in five different ways is 45. 135 can be written in 7 different ways; is this the smallest number that can be so written? What is the smallest number that can be written in 4 different ways? in 6 different ways?

An approach from the opposite direction

10. To express an odd number as the sum of two consecutive integers, we divide it by 2, and write the integer part (i) of the quotient followed by this number plus one:

$$111 \div 2 = 55; i = 55; 111 = i + (i + 1) = 55 + 56.$$

11. To express a multiple of 3 as the sum of consecutive integers, divide it by 3 and obtain q, and write $q - 1$, q and $q + 1$:

$111 \div 3 = 37; q = 37; 111 = 36 + 37 + 38.$

12. We do likewise with any number which has an odd factor: divide this factor and write that odd number of consecutive integers with the quotient in the middle.

$75 \div 3 = 25; 75 = 24 + 25 + 26,$

$75 \div 5 = 15; 75 = 13 + 14 + 15 + 16 + 17.$

13. If the number is even, divide by 4. If the remainder is 2 and the quotient is q, write $q - 1, q, q + 1, q + 2$.

$66 \div 4 = 16$ remainder 2; $66 = 15 + 16 + 17 + 18.$

14. If the number is an odd multiple of 3, divide by 6 to find the quotient and write $q - 2, q - 1, q, q + 1, q + 2, q + 3$.

$96 \div 6 = 16; 99 = 14 + 15 + 16 + 17 + 18 + 19.$

Next step

We can now generalise again.

15. Referring back to 9. above, if a number is an odd multiple of n, divide by $2n$ and write $q - (n - 1), \dots q, q + 1, \dots q + n$.

$119 = 7 \times 17$ $(n = 7)$: $119 \div 14 = 8$ remainder 7:

$119 = 2 + 3 + 4 + 5 + 6 + 7 + 8 + 9 + 10 + 11 + 12 + 13 + 14 + 15.$

But this rule does not work for $119 = 17 \times 7$ $(n = 17)$; $119 \div 34 = 3$ remainder 7: - or does it? Investigate. Suppose a number is an even multiple of n. Can we find a generalisation for this case?

16. We conjecture that all prime numbers, with the exception of 2 can be written as the sum of exactly two consecutive integers. Does the generalisation in 15 above apply to prime numbers. If so, does the generalisation help prove this conjecture?

Short cuts to adding

The numbers in the sum $5 + 6 + 7 + 8 + 9$ are in series and form a simple arithmetic progression, whose first term is 5 and whose successive terms differ by 1. We can calculate their sum (35) in various ways, bearing in mind how we may have created the sequence in the first place (for instance, see 15. above). The median number is 7: the number of terms is 5, and the sum is 5×7. Also both 5 and 9, and 6 and 8, add to give 14; the sum of the five numbers is

$$\frac{5}{2} \times 14 \text{ or } \frac{14}{2} \times 5.$$

Here we have an introduction to one of the general rules for finding the sum (S) of the terms of an arithmetic progression. We add the first term (a) and the last term (1), multiply this sum by the number of terms (n) and divide the product by 2: $S = n(a + 1)/2$. This second rule works when there is no single median number, as in $6 + 7 + 8 + 9$: the sum here is $(4 \times \{6 + 9\})/2 = 2 \times 15 = 30$.

Mechanising the process

Can you use a spreadsheet or write a computer program to find all the ways for writing a particular number as the sum of consecutive positive integers?

Tabulating

We can build a table to show the different ways in which numbers up to, say, 50 can be made by adding consecutive integers. Here the left-hand column indicates the first integer used, while the top row shows the number of integers used.

For example, 18 can be made from 3 integers of which the first is 5: $5 + 6 + 7$; or from 4 integers of which the first is 3: $3 + 4 + 5 + 6$.

	2	3	4	5	6	7	8	9
1	3	6	10	15	21	28	36	45
2	5	9	14	20	27	35	44	
3	7	12	18	25	33	42		
4	9	15	22	30	39	49		
5	11	18	26	35	45			
6	13	21	30	40				
7	15	24	34	45				
8	17	27	38					
9	19	30	42					
10	21	33	46					
11	23	36	50					
12	25	39						
13	27	42						
14	29	45						
15	31	48						
16	33							

It will be seen that column 2 contains the odd numbers, that is, numbers which are congruent to 1, modulo 2; column 3 contains multiples of 3; column 4 contains numbers which are congruent to 2, modulo 4; and that in general each odd-numbered column $(2n - 1)$ contains multiples of $2n - 1$ while each even-numbered column $(2n)$ contains numbers which are congruent to n, modulo $2n$.

A graphic view

Pupils may previously have met a type of figurate number called a trapezial number, in which each row in the figure contains one more symbol than does the row above it. Here is the trapezial number 24:

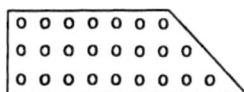

On inspection, this of course turns out to be no less than the solution to the problem of writing 24 as the sum of consecutive positive integers. But more than this, it also gives us a clue as to how to generate trapezial numbers, by completing the triangle of which the trapezium is the lower part:

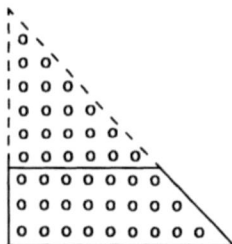

We see that the trapezial number 24 is the difference between the two triangular numbers

$$T_9 = \frac{1}{2} \times 9 \times 10 \text{ and } T_6 = \frac{1}{2} \times 6 \times 7$$

or $45 - 21$; and from this we go on to postulate that any trapezial number will be the difference of two triangular numbers.

Another table

We can generate the possible ways of writing numbers as the sum of consecutive positive integers by drawing up a table to show the differences between two triangular numbers.

	T_1	T_2	T_3	T_4	T_5	T_6	T_7
T_1	0	2	5	9	14	20	27
T_2		0	3	7	12	18	25
T_3			0	4	9	15	22
T_4				0	5	11	18
T_5					0	6	13
T_6						0	7

These are the same numbers which appeared in the previous table, arranged slightly differently. Study the patterns in the two tables. How do they differ? Can you extend the tables by using the patterns in each row or column?

The sequence of numbers (4), 9, 15, 22, 30,.... can be generated by the quadratic function

$$f(n) = \frac{1}{2}(n + 1)(n + 8).$$

Investigate further.

An analogy

We can write and calculate the sum of, say, $5 + 6 + 7 + 8 + 9$ as $T_9 - T_4$. This bears comparison with the method of finding the value of a product of consecutive integers, by which we calculate, say, $5 \times 6 \times 7 \times 8 \times 9$ as $9!/4!$ The value of $T_9 - T_4$ is

$$\frac{1}{2}(9 \times 10 - 4 \times 5) = 35.$$

Further considerations

Consider the following arguments. We have said already that 135 can be written as the sum of consecutive integers in seven different ways. To find these seven ways we write 135 as the product of pairs of all its factors:

$$1 \times 135; 3 \times 45; 5 \times 27; 9 \times 15.$$

This gives us immediately:

$3 \times 45, 44 + 45 + 46;$
$5 \times 27, 25 + 26 + 27 + 28 + 29;$
$9 \times 15, 11 + 12 + 13 + 14 + 15 + 16 + 17 + 18 + 19;$
$15 \times 9, 2 + 3 + 4 + 5 + 6 + 7 + 8 + 9 + 10 + 11 + 12 + 13 + 14 + 15 + 16.$

[Note that we cannot write 27 consecutive integers centred on 5 or 45 centred on 3 or 135 centred on 1 without using negative integers, although in each case these would add to give 135 – check this!]

We now double the first factor in each pair and halve the second:

$$2 \times 67\frac{1}{2}; 6 \times 22\frac{1}{2}; 10 \times 13\frac{1}{2}; 18 \times 7\frac{1}{2}.$$

This gives us three more ways; from $2 \times 67\frac{1}{2}$ we obtain 2 integers with a median of $67\frac{1}{2}$, which are $67 + 68$; also, $20 + 21 + 22 + 23 + 24 + 25$ and $9 + 10 + 11 + 12 + 13 + 14 + 15 + 16 + 17 + 18$, with median values of $22\frac{1}{2}$, and $13\frac{1}{2}$ respectively. We may not write 18 consecutive integers with a median value of $7\frac{1}{2}$ without using negative integers, although it does no harm to check that $-1 + 0 + 1 + 2 + 3 + 4 + 5 + 6 + 7 + 8 + 9 + 10 + 11 + 12 + 13 + 14 + 15 + 16$ do add to give 135. Check that the same would be true of 30 consecutive integers with a median of $4\frac{1}{2}$, of 54 with a median of 2, of 90 consecutive integers with a median of $1\frac{1}{2}$ and 270 with a median of $\frac{1}{2}$.

Use the same method to find the five ways of making 45 and the four ways of making 81. (Note that 81 is a square number: do square numbers show any special features?)

We are now in a position to suggest two more rules. Looking at 9 (2 ways), 18 (2 ways), 25 (2 ways), 36 (2 ways), 81 (4 ways), 162 (4 ways), we see that it appears that we can write square numbers and multiples of square numbers, where the multiple is a power of 2, in an even number of different ways, unless the square number is itself a power of 2, for example, 4 or 64, in which case it apparently cannot be written at all as the sum of consecutive integers. All other numbers can be written only in an odd number of ways.

The number of ways in which we can write any number as the sum of consecutive integers depends on the number of odd factors it possesses and on the comparative size of each odd factor and its complementary factor. For example, 63 has 6 odd factors: 1, 3, 7, 9, 21 and 63. Complementary pairs are 1 and 63, 3 and 21, and 7 and 9. Because half of 7 is less than 9, we can have 7 integers centred on 9; and because half of 9 is less than 7, we can have 9 integers centred on 7; but although we may have 3 integers centred on 21, we cannot have 21 (positive) integers centred on 3 because half of 21 is not less than 3.

We then double each odd factor and halve its complement: 2 and 31½, 6 and 10½, 14 and 4½, 18 and 3½, 42 and 1½, and 126 and ½. Because half of 2 is less than 31½, and because half of 6 is less than 10, we have 31 + 32 and 8 + 9 + 10 + 11 + 12 + 13 as additional ways for 63, but this is all, since, for example, we cannot find 14 positive integers with 4½ as the median value since half of 14 is greater than 4, or to put it another way, 7 is greater than half of 9.

We now apply these rules to 315.

$$315 = \quad \begin{array}{ll} \mathbf{1 \times 315} & \mathbf{2 \times 157\tfrac{1}{2}} \\ \mathbf{3 \times 105} & \mathbf{6 \times 52\tfrac{1}{2}} \\ \mathbf{5 \times 63} & \mathbf{10 \times 31\tfrac{1}{2}} \\ \mathbf{7 \times 45} & \mathbf{14 \times 22\tfrac{1}{2}} \\ \mathbf{9 \times 35} & \mathbf{18 \times 17\tfrac{1}{2}} \\ \mathbf{15 \times 21} & 30 \times 10\tfrac{1}{2} \\ 21 \times 15 & 42 \times 7\tfrac{1}{2} \\ 35 \times 9 & \end{array}$$

In the left-hand column any combination (in bold) in which half an odd factor is less than the complementary factor will give us a solution: in the right-hand column those possibilities will give a solution where half the even factor is less than the other factor; and so 315 can be written as the sum of consecutive integers in the 11 ways indicated.

Powers of 2

We noted earlier that we do not appear to be able to write powers of 2 as the sum of consecutive positive integers. Suppose we could do so for some power of 2, say 2^{k-1}, so that $T_m - T_n = 2^{k-1}$. Then,

$$\frac{1}{2}m(m+1) - \frac{1}{2}n(n+1) = 2^{k-1}$$

$$m(m+1) - n(n+1) = 2^k, \text{ doubling both sides.}$$

$$(m^2 - n^2) + (m - n) = 2^k$$

$$(m - n)(m + n + 1) = 2^k.$$

Now if m and n are both even, then $(m - n)$ is even and $(m + n + 1)$ is odd, so the left-hand side has an odd factor. However the right-hand side does not have an odd factor, all its factors being powers of 2, which are all even. So m and n cannot be both even.

240

If m and n are both odd, then $(m - n)$ is again even and $(m + n + 1)$ is again odd, so the same argument applies.

If either of m and n is even and the other is odd, then although $(m + n + 1)$ is now even, $(m - n)$ is now odd, so the argument applies again. There are no other possibilities for the parities of m and n, so we conclude we cannot find m and n such that $T_m - T_n = 2^{k-1}$, and that our attempts to express, say, 16 or 64 as the sum of consecutive integers are doomed to failure. Are you convinced that we have proved this is so? Suppose we are allowed to use negative integers? What then? Investigate!

6.8 Roger Porkess on '60° triples'

Vol. 27, no. 1 (January 1998), pp 10-12

Introduction

From time to time many teachers set their students investigations into Pythagorean Triples. A similar but less familiar situation arises with triangles which have one angle 60° and integer length sides, like that drawn below.

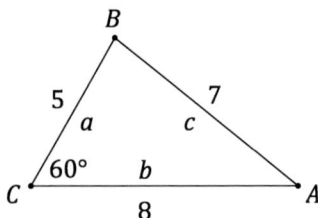

> How can you generate the complete set of triples $\{a, b, c\}$ which give the sides of such triangles?

In this article the solution to this problem is set out in a number of stages following the general route which a typical student might follow. According to the depth to which it is taken, it is suitable for use with good Higher Tier GCSE students or with those taking A Level.

Solving the problem

Stage 1: Look for patterns

Students will almost certainly start by looking for other triples in the hope that when they find some, a pattern will emerge. This is unlikely to prove a successful strategy. With the exception of equilateral triangles, the triples are not easy to find; nor do they give away much information when they are found. The problem needs to be looked at algebraically.

Stage 2: Using algebra

Applying the cosine rule to the triangle in the diagram:

$$c^2 = a^2 + b^2 - 2ab \cos 60°$$

and so $c^2 = a^2 + b^2 - ab.$

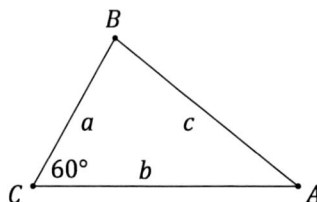

The next few steps, leading to equation (1), are ones which most students will not take naturally. This is not so much a case of their not being able to do it as of their not recognising the significance of the result.
Rearranging $c^2 = a^2 + b^2 - ab$ gives

$$c^2 = a^2 - 2ab + b^2 + ab$$
$$c^2 = (a - b)^2 + ab$$
$$c^2 - (a - b)^2 = ab$$
$$(c - a + b)(c + a - b) = ab \qquad\qquad (1)$$

The process of solving equation (1) is presented in three stages of progressively increasing sophistication.

Stage 3: First attempt at solving equation (1)

An obvious solution is obtained by setting one of the factors on the left hand side equal to a and the other equal to b.

$$\begin{cases} c - a + b = a \\ c + a - b = b \end{cases}$$

This gives $a = b = c$, an equilateral triangle. Equating the factors to b and a respectively gives the same result.

Clearly we have achieved partial success. The equilateral triangle is a valid solution but certainly not the only one; the triangle with sides 5, 8 and 7 in the first figure is another.

Stage 4: A second attempt at solving equation (1)

A possible next step is to set

$$\begin{cases} c - a + b = an \\ c + a - b = \dfrac{b}{n} \end{cases}$$

where n is an integer ≥ 1.

Eliminating c between these equations gives

$$b = a\left(\frac{n^2 + 2n}{2n + 1}\right).$$

Since a and b are integers it is reasonable to let $a = 2n + 1$. In that case it follows that $b = n^2 + 2n$ and $c = n^2 + n + 1$.

Note that if a is taken to be $p(2n + 1)$ for some number p, then $b = p(n^2 + 2n)$ and $c = p(n^2 + 2n + 1)$. In other words it is the same solution multiplied through by p. Throughout this article triples which are multiples of each other, for example $\{9, 24, 21\}$ and $\{3, 8, 7\}$, are treated as the same.

At this stage it is advisable for students to check that if $a = 2n + 1, b = n^2 + 2n$ and $c = n^2 + n + 1$ then C = 60°. (It does!)

Substituting successive values of n gives this table.

n	$a = 2n + 1$	$b = n^2 + 2n$	$c = n^2 + n + 1$	Simplified
1	3	3	3	1, 1, 1
2	5	8	7	
3	7	15	13	
4	9	24	21	3, 8, 7
5	11	35	31	
6	13	48	43	
7	15	63	57	5, 21, 19
8	17	80	73	
9	19	99	91	
10	21	120	111	7, 40, 37
11	23	143	133	
12	25	168	157	

Stage 5: Is this the complete set?

Clearly we now have a means of generating as many triangles as we wish. The question is, "Have we caught them all?" The answer is "no" and a missing triple is not too hard to find.

Looking at the numbers in Table 1 reveals two rather similar sets

$$n = 2: \{5, 8, 7\},$$

$$n = 4: \{3, 8, 7\}.$$

The explanation for such a pair can be seen in the next figure. The line CB in the original $\{5, 8, 7\}$ triangle ABC has been extended to D to form an equilateral triangle ACD. The triangle DAB also has integer sides; their lengths are 3, 7 and 8 and the angle D is 60°.

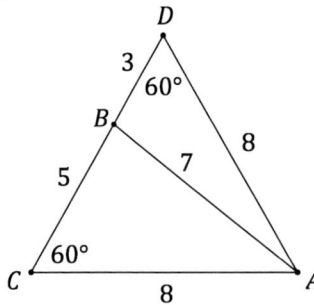

The same procedure can clearly be applied to every other (non-equilateral) triangle of this form to give a conjugate triple. Thus $\{7, 15, 13\}$ has a conjugate $\{8, 15, 13\}$ since $15 - 7 = 8$.

However, the triple $\{8, 15, 13\}$ does not appear in the table and so we conclude that there are solutions other than those generated by the present procedure. We must look again at our procedure for solving equation 1 and seek to make it more general.

Stage 6: An even more sophisticated solution to equation (1)

The existing procedure for solving equation (1) involves setting

$$\begin{cases} c - a + b = an \\ c + a - b = \dfrac{b}{n} \end{cases}$$

What happens if instead of the integer n, we use the rational numbers n/m? We now have

$$\begin{cases} c - a + b = \dfrac{n}{m}a \\ c + a - b = \dfrac{m}{n}b \end{cases}$$

and, using the same procedure as before, these give a solution

$$a = m^2 + 2mn$$
$$b = 2mn + n^2$$
$$c = m^2 + mn + n^2.$$

The first table (above) was constructed using $m = 1$. Other values of m give different tables; those for $m = 2$ and $m = 3$ are given below.

n	$a = m^2 + 2mn$	$b = 2mn + n^2$	$c = m^2 + mn + n^2$	Simplified
1	8	5	7	
2	12	12	12	1, 1, 1
3	16	21	19	
4	20	32	28	5, 8, 7
5	24	45	39	8, 15, 13
6	28	60	52	7, 15, 13

Notice that the triple missing from the first table, $\{8, 15, 13\}$, has now turned up under $m = 2, n = 5$.

n	$a = m^2 + 2mn$	$b = 2mn + n^2$	$c = m^2 + mn + n^2$	Simplified
1	15	7	13	
2	21	16	19	
3	27	27	27	1, 1, 1
4	33	40	37	
5	39	55	49	

Stage 7: Tidying up

Have we now covered all possibilities? We looked for solutions to equation (1) of the form

$$\begin{cases} c - a + b = \dfrac{n}{m}a \\ c + a - b = \dfrac{m}{n}b \end{cases}$$

This involved three assumptions, all of which are easily justified.

Assumption 1 That the number multiplying a must be rational (i.e. of the form n/m.)

Justification Write $c - a + b - qa$. Then $q = \dfrac{c - a + b}{a}$ which is rational since a, b and c are all integers. So q must be expressible in the form n/m.

Assumption 2 That this multiplying number must also be positive.

Justification The number q must also be positive since the triangle of inequality requires that $b + c > a$.

Assumption 3 That the forms of the right hand sides were the given way round, i.e. a with the first equation and b with the second.

Justification Suppose instead we write

$$c - a + b = \frac{m}{n}b \text{ and } c + a - b = \frac{n}{m}a .$$

These relationships can be rewritten as

$$c - a + b = \left(\frac{mb}{na}\right)a$$

$$c + a - b = \left(\frac{na}{mb}\right)b$$

which is the same form as before but with mb replacing n and na replacing m.

Stage 8: Checking

Have we now really caught all the triples? If we have really considered all the conditions attached to the argument, then the answer must be yes. However, under stage 7 we justified 3 assumptions in the argument; are there others we have not considered?

It is hard to feel secure in this sort of proof. You can check it time and again and always miss the same point. In such a situation it is helpful to have an independent check and this can be provided by a computer printout of triples. In a suitable program a, b and c might run through all integer values between 1 and 1000 and be printed out in those cases where:

$$a^2 + b^2 - ab - c^2 = 0.$$

You then have to check that each of the triples in the print out corresponds to particular values of m and n.

This raises other questions. How can you test that there are values of m and n which generate $\{561, 840, 741\}$ or any other given triple? The triple in question reduces to $\{187, 280, 247\}$; what about that case? How can you construct an algorithm so that the computer can do the work of finding m and n in each case? (Otherwise it will take you ages to check them all.) The strength of using the computer triples is that they are generated by a completely different method. However, students should realise that checking each one up to 1,000, or even 1,000,000, does not constitute a proof that the earlier work is correct.

Related problems

This problem provides a starting point for other lines of investigation including:

i.   Use a similar method to derive the Pythagorean Triples generator $\{m^2 + n^2,\ m^2 - n^2, 2mn\ \}$.

ii.   Find an expression in terms of m and n for the smallest side, a', of the conjugate triangle. Use a spreadsheet to calculate the angles A and B in the triangles generated when $m = 1$ and $1 \le n \le 40$, and the angles A' and B' in their conjugates. The graph below, which was drawn using a spreadsheet, shows A, B, A', B' for $m = 1$ and $1 \le n \le 40$. What happens for other values of m?

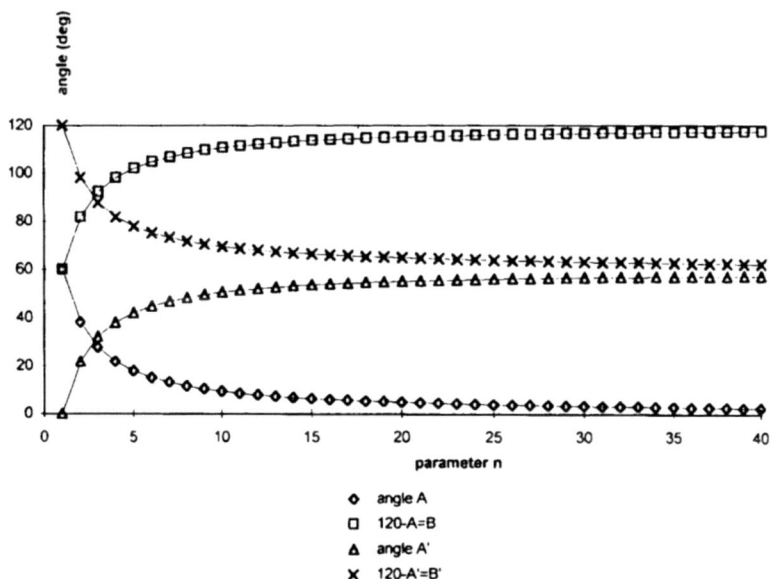

iii.   Can you find triangles with integer sides and a given angle which is neither 60° nor 90°?

Conclusion

Much concern has been expressed recently about students' lack of both algebra skills and understanding of proof. This piece of work addresses both within an essentially investigative setting; students who work through it stand to learn a great deal more than which particular triangles fit the given requirements.

Acknowledgements

The Editors wish to thank all the authors of the articles selected for this compilation. They have given us a wealth of material to try in the classroom, material that will interest, even inspire, and certainly challenge.

The book would not have come together without the leadership of Jill Trinder, the Chair of the Mathematical Association's Publications Committee, the typesetting skills of Bill Richardson and the administration assistance provided by Amber Richardson. We are grateful for their input.

Finally, thanks to our team of proof-readers, Ian Anderson, Sue Childs and Ray Huntley. They spotted errors and typos that others simply would not have seen and of course that makes for a better product for you, the reader.

Chris Pritchard & John Berry